Clinical Sociology: Research and Practice

Series Editor
Jan Marie Fritz
University of Cincinnati, Cincinnati, Ohio, USA
University of Johannesburg, Johannesburg, South Africa

The series explores key research and practice in this rapidly expanding field. Topics include ethical and legal aspects of intervention; the nature of client relationships; methods of intervention and evaluation; and the role of clinical sociology in specific settings. This open-ended series appeals to practitioners, policymakers, students and professors in sociology, social work, community psychology, public health, health education, social policy and counseling.

Vidette Bester

The Untold Story of Zama Zama Miners in South Africa

Unearthing Hope

Springer

Vidette Bester
Research Associate, Sociology Department
University of Johannesburg
Johannesburg, South Africa

ISSN 1566-7847
Clinical Sociology: Research and Practice
ISBN 978-3-031-82782-2 ISBN 978-3-031-82783-9 (eBook)
https://doi.org/10.1007/978-3-031-82783-9

This Springer imprint is published by the registered company Springer Nature Switzerland AG
The registered company address is: Gewerbestrasse 11, 6330 Cham, Switzerland

If a story moves you, act on it.

Sisonke Msimang

Preface

At the time of writing this, the situation surrounding illegal miners in Stilfontein, South Africa, has escalated dramatically. An estimated 400–4000 miners are believed to be trapped in a disused mine shaft, facing dire conditions without access to food or water. The South African government has adopted a strict stance against illegal mining, choosing not to assist these miners as part of its broader strategy to combat the issue. Community members have taken it upon themselves to attempt rescues, leading to a tense standoff with authorities. A recent court ruling mandated that the mine shaft be unblocked to allow trapped miners to exit safely while police continue to monitor the situation closely, emphasizing that those who resurface will most likely be detained. It is important to note that this situation is ongoing, and the outcome is uncertain at this stage, but it looks grim for all parties involved.

The Stilfontein situation should be considered in the broader context. In 2023, President Cyril Ramaphosa authorized the deployment of 3300 South African National Defence Force (SANDF) personnel to assist the South African Police Service in combating illegal mining activities. The reasons for this crackdown are valid. Alarming incidents such as the gang rapes of women in 2022, attributed to illegal miners, and two tragic gas explosions—one in May 2023 in Welkom that killed 31 miners, and another in June 2023 in Boksburg that claimed numerous lives—are behind the South African government's stance on illegal mining.

This, of course, is one side of a complex story.

The other side is that desperation and poverty make many Zama Zamas engage in artisanal mining. As highlighted in their accounts and presented in this book, this path paradoxically offers a way to make a living without resorting to crime for many men. For the women involved, it provides an escape from exploitative domestic work, serving as a means of economic empowerment, as they shared with me.

The Zama Zama issue is so complex because it has many faces—each face tells a different story. I hope this book sheds light on the diversity of the sector and the many faces, so to speak. Acknowledging its dark and gray-faced sides can help us step back to see all faces simultaneously. This may allow us to reject the simplistic story of violence, criminality, and illegal immigrants, and as Chimamanda Ngozi Adichie said: *"When we reject the single story when we realize that there is never a single story about any place, we regain a kind of paradise."*

Johannesburg, South Africa Vidette Bester

Ethics Approval

All interviewees in this book gave informed consent to participate in the study. Direct quotations from the interviews are presented verbatim throughout the book, and pseudonyms are used.

Contents

About the Author

Vidette Bester has made significant contributions to the field of artisanal mining in South Africa. With a doctorate in Sociology, her work explores the socioeconomic drivers of artisanal mining, the role of mining companies in the sector, and the impact on miners' health and safety. She has delved into gender dynamics within artisanal mining, highlighting the unique challenges faced by women in this sector. Dr. Bester is an advisor to mining companies, and she has developed a comprehensive framework to address artisanal mining in the South African context, aiming to improve working conditions, safety, and the socioeconomic benefits for historically disadvantaged communities. Her work advocates for more sustainable and proactive approaches to address the issues associated with artisanal mining rather than relying solely on policing and criminalization.

List of Figures

List of Tables

Chapter 1
Re-thinking Zama Zama Mining

It's the role of narrative to bridge the gap between philosophy as an abstract theory, ideas in the ether, and life as lived on the ground.

—Held (2016)

Introduction

My journey into the world of artisanal mining commenced in 2011 when I embarked on a research expedition to the southern region of the Democratic Republic of the Congo (DRC). At that time, I was working as a researcher at a research unit affiliated with one of South Africa's universities. This expedition showed me the deep poverty levels and desperation of the artisanal miners I encountered, a sight I will never forget. The trip served as a stark reminder of the prevailing power imbalances in developing nations, with governments and corporations on one side and the impoverished on the other. Upon my return to South Africa, the stories of artisanal miners—men, women, and children alike—remained etched in my memory.

The issue of artisanal mining, also known as "Zama Zama mining," entered the South African mainstream media a few years after I visited the DRC. Zama Zama is an isiZulu phrase meaning "we are trying." Often, Zama Zama mining is described or labeled as illegal mining. Consequently, media coverage has primarily portrayed Zama Zama miners as violent, criminal men who have illegally immigrated to South Africa. The response from the South African government and mining houses has involved policing the sector, resulting in the further marginalization of this group. My experience in the DRC urged me to look at this phenomenon differently.

As a result, I enrolled in a PhD program in Sociology at the University of Johannesburg in 2017, aiming to delve deeper into this issue within the South African context. The central research question of my thesis revolved around the socioeconomic development of artisanal mining in South Africa. I had more questions than answers at the end of my PhD journey. This time, my inquiries expanded to encompass pressing issues related to poverty, historical legacies, and power dynamics. I also pondered issues about the artisanal mining sector and the broader

V. Bester, *The Untold Story of Zama Zama Miners in South Africa*, Clinical Sociology: Research and Practice, https://doi.org/10.1007/978-3-031-82783-9_1

socioeconomic systemic conditions, questions around women in artisanal mining, and how the South African mining history, specifically the migrant labor system, has shaped the country's current Zama Zama mining sector. With these questions in mind, I started writing this book, hoping to understand an extraordinarily complex and fluid phenomenon better.

As I began writing this book, I realized that although it is about artisanal mining in the South African context, it is also about language, poverty, and power. By reshaping and reframing the artisanal mining debate in South Africa, the book may also explore how we label and contribute to the dominant societal narrative—and exclude and marginalize vulnerable socioeconomic groups. As members of society, we have become blinded by labels and their impact on people and groups. It divides and marginalizes vulnerable socioeconomic groups but also affects how we see, describe, and label the world. As a result, we have become players in the cycle of labeling, exclusion, and continuously etching the socioeconomic fault lines deeper.

This acknowledgment poses the question of who has the power to label and shape the mainstream discourse. In his work on classification, the French sociologist Pierre Bourdieu (2019) argues that labeling and classifying things informs how we view things. According to Bourdieu's work, we continuously make assumptions about the class in which we categorize the person we engage.

Webster, Lambert and Bezuidenhout (2008) explain that their research participants experienced individual insecurity in their book *Grounding Globalization*. The authors refer to C Wright Mills's work and his view of a "*personal trouble.*" Webster et al. (2008) explain that: "*when large numbers of workers in different factories in different countries experience the same feelings of insecurity, it is no longer a personal trouble only; it is a public issue.*"

The same line of thought can be applied to the matter of the artisanal miners. If several people in a marginalized socioeconomic group like the artisanal miners experience insecurity, exclusion, and poverty as they do, then surely, we must take a step back to examine the broader context around this, since it is a public issue.

However, dissecting the aspects of a socioeconomic landscape calls for a journey into our modern world's systemic socioeconomic fault lines. It also asks us to look back into history. More importantly, it is not enough "*…to identify the sense of helplessness, fear, desperation, anger, and sadness, often self-destructive behavior such as suicide, substance abuse, and domestic violence*" (Webster et al., 2008). We must take a step or two back and put on a new set of lenses to try to begin to understand how insecurity, marginalization, and exclusion are being created within society. As Mills says, what is needed is to identify the broader social forces, institutions, and organizations that generate this insecurity (as cited in Webster et al., 2008).

After more than a decade's research and writing on artisanal mining, I also realized that it is nearly impossible to understand artisanal mining without considering the broader historical context of South Africa's mining history. For one, the notorious mining labor system introduces the notion that the mining industry has a moral obligation regarding the social ills it helped to create. Such an acknowledgment also explores how this notorious system shaped the course of artisanal mining. Since South Africa's mining industry was built on the backbone of migrant labor, it is worth looking at how migration played into the artisanal mining context. After all, history does matter. Only by taking a closer look at the questions I have mentioned on the previous pages will we hopefully gain a more reflective understanding of this sector, and only through the voices of the artisanal miners will we be able to break down some preconceived perceptions and ideas. Such a stance, of course, allows for entertaining more than one narrative. As Chimamanda Ngozi Adichie (2009) said:

> Many stories matter. Stories have been used to dispossess and to malign. But stories can also be used to empower and to humanize. Stories can break the dignity of a people. But stories can also repair that broken dignity.

This chapter's subsequent parts unpack how narratives and language can marginalize individuals, particularly those with less societal power. Following this discussion, I will share my approach as a Clinical Sociologist to the issue of Zama Zama mining in South Africa. Furthermore, the chapter explores how the media contributes to a broader, often misguided discourse on the Zama Zamas, followed by examining the prevalent mainstream labels and misconceptions surrounding Zama Zama miners.

Social Exclusion and the Power of Words

Much has been written about the interplay between power, the creation of narratives, and knowledge formation in sociology and other related disciplines. Being in a powerful societal position almost always grants one the privilege to influence the broader discourse about some issue or even in general. Societal structures like governments and corporations, or anyone or anything with a highly invested socioeconomic interest and big enough power base, are most likely in a position to shape the societal discourse and dominant ideologies.

The powerful imprints on our psyche, often beyond our awareness or consent, shape the filters, words, and labels we apply in life. Certain groups have more say than others during the interplay between power and the creation of narratives. On the other hand, specific groups get disqualified from participating in developing knowledge or a dominant ideology. Their voices are excluded in the central space where significant political, social, or economic decisions are made. The reasons for the disqualification can vary. It can be race, gender, sexual orientation, or class, to mention but a few. The list can go on because, as in the words of Sidel Germeten:

> …some stories are kept in silence by 'others', and some stories are related by loud voices. There are also groups of people who have not been able or permitted to tell their life stories to society. Historically, these groups have included women, indigenous people, and persons classified as 'mentally retarded'. (Germeten, 2013)

These words by Germeten can justifiably be applied to any other group that has been oppressed and subverted into an inferior position in society through having no voice or, at its best, being allowed a limited space in a narrative. As sociologist Alex Honneth (1991) notes, exclusion from society ultimately denies a person the recognition needed to be included. Furthermore, from this perspective, poverty can be regarded as a lack of material resources and a significant deficit in social recognition (Honneth, 1996, 2004, 2014). Honneth (1996) argues that poverty stems from and contributes to a lack of recognition, leading to social exclusion, shame, and stigmatization. This lack of recognition affects individuals' self-esteem and ability to participate fully in society. Addressing poverty requires more than just economic measures; it also necessitates fostering social esteem and respect for marginalized individuals and communities (Honneth, 1996, 2014).

As Michel Foucault (see Foucault, 1972, 1981) theorized years ago, vocabulary, language, and concepts play a role in our everyday lives. For Foucault, discourse refers to a historically conditional social system that generates knowledge and meaning. Discourse is a mode of organizing knowledge that structures the formation of social (and increasingly global) relations through collective comprehension of discursive logic and acceptance of discourse as a social fact. According to Foucault, the logic produced by a discourse is structurally related to the broader structure of knowledge of the historical period in which it emerges (Foucault, 1972). Discourses are produced by the effects of power within a social order, and this power prescribes specific rules and categories that define the criteria (Foucault, 1972, 1981). Foucault asserted that naming and classifying are dangerous social games. He asked who has the authority to articulate, modify, circulate, or employ the statements and schemes they contain (Snyder, 1984). Discourses can be a powerful tool to restrict alternative thinking or expressing ideas; therefore, knowledge becomes a force of control (Giddens, 2001).

Regarding artisanal mining, Foucault's words are far from dead or irrelevant. In 2015, the South African Human Rights Commission (SAHRC, 2015) called for a comprehensive framework to strengthen the profile of the Zama Zama miners and improve artisanal miners' practices. The SAHRC rightly anticipated that the one-dimensional illegal label would only exacerbate the social impacts surrounding this sector. A decade after the SAHRC called for a more nuanced version of the artisanal miners' profile, the Zama Zama miners are still mostly reduced to a simplified description of men who illegally immigrated to South Africa and have a propensity for violence and crime. Unfortunately, this lack of more profound knowledge feeds the cycle of classification, labeling, and, in the end, social exclusion. During one discussion, a civil society worker shared his view, which captures this cycle well:

> "Yes, the Department of Mineral Resources, because they should be engaging. And the biggest problem is the DMR doesn't want to open a platform and talk to them because they say they are not prepared to talk to criminals."—Mike, civil society

Inevitably, by being labeled illegal, the cycle of exclusion gets fueled. Moreover, as long as social exclusion remains in place, the label of illegality persists, and poverty prevails.

More Than a Single Story: Artisanal Mining Through the Clinical Sociology Lens

One way to reconsider the matter of artisanal mining is through the theoretical viewpoints of clinical sociology.

Clinical sociology is a creative, rights-based, and interdisciplinary specialization that seeks to improve life situations for individuals and the collective (Fritz, 2008). A rights-based approach means that a clinical analysis and intervention aims to "*...promote and maintain a minimum standard of well-being to which all people would ideally possess a right*" (Johnson & Forsyth, 2002). Hall (2006) also describes the pragmatical side of clinical sociology by stating that "*...clinical sociology focuses on designing effective intervention strategies and articulating productive didactic between professional clinical sociologists and their clients.*"

In other words, clinical sociologists create or alter existing systems to assist in change and assessment. Furthermore, clinical sociology allows or produces sociological knowledge to understand a situation better and critically (Rhéaume, 2014). Clinical sociologists are scientists who are humanistic and multidisciplinary in their approach. They practice in diverse settings at all levels of social organization (Bruhn & Rebach, 1996). In addition, clinical sociologists engage in planned social change by focusing on one system level (interpersonal, community, international) but integrate levels of focus in their work and do so from a sociological frame of reference (Fritz, 1989).

By ensuring social change, creativity plays an integral part in the field of clinical sociology. Since clinical sociology emphasizes interventions and how community development or interventionist strategies can take place, a creative approach to such work is much needed. Creativity must be innovative (Bilton, 2007). Creativity is also a process and an outcome and refers to "*... ideas of duality and paradox, the combination of ideas into new and unexpected patterns, combinations of innovation and value, of different thinking styles, rationality, and irrationality*" (Bilton, 2007).

Clinical sociologists use various methods in their research, but an emphasis on qualitative approaches exists. Examples of qualitative approaches include, but are not limited to, participatory, collaborative, or life history and qualitative case-study methods (Fritz, 2013). As a result, clinical sociology allows or produces sociological knowledge to reach a better and more critical understanding of a situation (Rhéaume, 2014). Community development is a significant area of social practice. It refers to people who, on a territorial basis, organize themselves to ensure a better quality of daily life and resist the dominant neoliberal ideology, which is reproducing and increasing social inequalities (Rhéaume, 2014).

The emergence of clinical sociology in South Africa is relatively recent. However, since sociology's inception in universities, much of the work could be viewed as belonging to this field (Uys, 2021). Uys (2021) elucidates how, in 1984, Webster advocated for a form of sociology that would adopt *"a critical perspective toward social structures by unveiling disparities, manifestations of power, distortions in ideologies, and their origins ...*" As Webster articulated, the 1980s represented a zenith for sociology in South Africa, marked by the emergence of "*a diverse and lively sociological community deeply engaged in conversations with emerging social movements that were vehemently opposing apartheid.*" This context allowed South African sociologists to draw nearer to the ethos of clinical sociology (Uys, 2021).

Examining artisanal mining within the South African context by using the theoretical principles of clinical sociology fosters creativity, leading to the development of new ideas and intervention strategies for the sector. With this book, the focus is on fostering social change for the artisanal miners, by integrating the voices of various stakeholders. Also, by contextualizing the artisanal mining situation within the history of the formal mining industry in South Africa paves the way for a deeper understanding and a fresh perspective on the sector. It establishes a standpoint from which the exploitative history of the formal mining industry cannot be ignored anymore because denying people access to land and mineral rights opens up the question of power and legacy issues.

Moreover, by juxtaposing the activities of the formal mining industry with those of the Zama Zama miners, we can provoke a thoughtful exploration that encourages a reevaluation of the societal and environmental impacts imposed by the formal mining sector. Paradoxically, this critical discussion often remains in the shadows, receiving minimal media attention. Engaging with the artisanal miners by exploring their experiences within the sector creates a more considerable and sociological understanding of the heterogeneous nature of people within the sector. By erroneously viewing the artisanal mining sector as a homogenous one, which is standard practice, one misses the rich diversity of the sector.

To create an alternative narrative, qualitative data must be gathered. Interviews were conducted with various stakeholders to uncover people's perceptions, views, and experiences regarding artisanal gold mining. Babbie and Mouton (2014) advised to consider the following criteria when selecting research participants: enculturation, current involvement, and the availability of adequate time. This means that research participants should be well-informed, active in the field, and willing to make sufficient time available to be interviewed for the research. Thus, to gain a well-informed, heterogeneous perspective of artisanal mining in South Africa, participants who have worked with, researched, or who have experience in dealing with artisanal miners were selected for interviews.

In addition, the selected participants included those who could provide critical insight into the phenomena. Purposive and snowball sampling were used to select the study participants. Data was collected in 2018 and 2023. The interviewees can broadly be categorized into four groups: Zama Zamas, civil society workers, representatives of mining companies, and key informants. Direct quotations from the interviews are presented verbatim throughout the book, and pseudonyms are used.

For this book, all interviews took place in and around the Witwatersrand area of South Africa. This area, known for its gold mining operations, is the breeding ground for many Zama Zama gold mining activities in South Africa. The SAHRC

(2015) confirmed that the Gauteng Province is most affected by Zama Zama gold mining activities, hence the focus on this province.

Zama Zama mining sites are often located next to, or even atop, abandoned gold mines. These locations overlap with informal settlements that are part of the surrounding former or current mining communities. Figure 1.1 illustrates how informal settlements are often situated in close proximity to gold mining operations.

Interviews were conducted with Zama Zamas, both men and women, at the artisanal mining sites, which contributed significantly to understanding who artisanal miners are. With these empirical perceptions, experiences, and understandings, I aim to portray the matter as more nuanced.

Most importantly, I hope this book underscores how the dominant current discourse around the artisanal mining sector has morphed into a narrow narrative. A homogenous view of the sector ignores that people differ in their life stories and experiences. Dropping loaded labels such as illegal, criminal, and violent opens new pathways to a more nuanced understanding of the sector. In many instances, labels pigeonhole Zama Zamas, reducing their multifaceted identities or experiences to a single, one-dimensional portrayal. This one-sided narrative can perpetuate stereotypes, misconceptions, and biases, fostering an incomplete understanding of the sector.

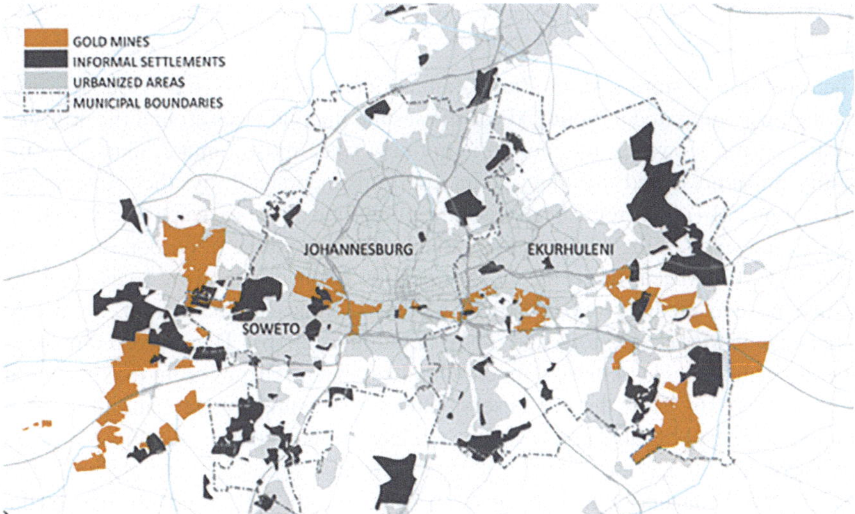

Fig. 1.1 Informal settlements and gold mines (Enviro Editor, 2018)

Zama Zamas and the South African Media

During the conversations, several interviewees stated that formal industries, like mining companies and the government, dominate the current narrative. Interviewees believed journalists automatically accept a narrative carefully choreographed by the government and formal mining sectors. One participant claimed that reporters avoid researching Zama Zama mining. During a conversation, a civil society worker mentioned that journalists would rather repeat stories about shootings.

> "Well, I think, first of all, the journalists are extremely lazy, and they don't want to make the time to meet with small-scale miners. They base their accounts on other media reports and those media reports are based on, it's all second-hand information. And it usually comes, the media usually writes about mining when there has been a shooting incident where syndicates have tried to take off the gold or the proceeds."—Danny, civil society

In conversation with a key informant, a rhetorical question was asked why the media scrutinizes the Zama Zama mining sector but not the mining houses.

> "And why are we asking questions around the artisanal miners and not around the mining houses? The reason is because the people asking the questions are the mine houses themselves. So that's the discourse or the narrative that they draw out, and it's a very misleading narrative."—Jack, key informant

A former journalist and currently a writer remarked that the media oversimplifies the complex issue of Zama Zama mining.

> "Having worked in the media, the media is very good at simplifying complex issues; this is a very complex issue."—Travis, key informant

The repetition of stories where Zama Zama miners are portrayed as violent, criminal men who migrated to South Africa misses the finer nuances around the artisanal mining sector. However, this notion is not new since most Zama Zamas are commonly misrepresented by the media and misunderstood by the government (Thornton, 2014). This reinforces misconceptions and creates a deeper skewed story.

This tendency of the media to favor the actions and nature of corporations is not new. In South Africa, the print media operates in an oligopolistic market and is owned by five media groups, namely Media24, Times Media Group, Sekunjalo Independent Media, Caxton, and TNA (Chiumbu, 2016). These five companies operate within a logic of market fundamentalism that privileges profit maximization (Chiumbu, 2016). How the media serves the mainstream economy, and the formal mining industry, became, once again, evident in the reporting of the Marikana massacre.

The Marikana massacre happened on August 16th, 2012, and resulted in the death of 34 striking Lonmin miners at the hands of the police. The massacre has been regarded by many as the worst act of violence by the South African police since the end of apartheid in 1994 (Rodny-Gumede, 2015). The run-up to the massacre had a long and complex history, with miners' requests for an increase in wages. How the media portrayed the events, pre- and post-the massacre, reflects how the media serves the interest of corporate power. Scholars explain that, broadly, the media created and endorsed an acceptable and uncritical discourse of Marikana. The media's discourse aimed to justify the liberal capitalistic system by not allowing any system critique (Duncan, 2014; Chiumbu, 2016). During the entire strike, the media focused on describing single acts of actual or potential violence by the miners. Generally, they underplayed the political goals and causes of the struggles (Duncan, 2014; Chiumbu, 2016). Duncan (2014) writes that the miners were described in the media as predisposed to violence and that they held superstitious beliefs. These narratives cut across the business and general press and were implicitly biased toward the police and business cases.

Furthermore, Duncan (2014) clarifies that much media reporting on Marikana shifted after independent accounts of the events had been published, leading to greater sensitivity. Indeed, the media published several stories that showed the human face of the miners and their families. Duncan emphasizes that the sympathetic articles on Marikana raised legitimate concerns regarding low wages, terrible living conditions, and police brutality. However, in most of these sympathetic stories, others spoke on behalf of the miners, raising issues they thought the miners were facing. The miners were not asked to give testimonies and stories about their realities.

In Chiumbu's research, which entails the analysis of 60 newspaper articles, there was a scattering of political commentators, politicians, activists, and business consultants who raised issues about the working and living conditions of the miners (Chiumbu, 2016). Also striking about these articles is that they are framed in a particular ideological logic and do not cross the boundaries by unsettling dominant discourses (Chiumbu, 2016).

The same line and flavor of reporting are used when mainstream media report on Zama Zama mining. Stories of violence and turf wars are often reported. Although these incidents do happen, what is lacking is the incorporation of the voices of the artisanal miners, men, and women.

As described, this tendency to disregard the heterogeneous nature of artisanal mining in the South African context has led to the Zama Zama sector generally being portrayed negatively. As an example, the media and mining representatives, such as the Minerals Council (2023) often report that the miners are from neighboring countries, such as Zimbabwe and Mozambique, labeling them as illegal or undocumented immigrants (Ryan, 2019; Seleka, 2019; Rademeyer, 2022; Mabena, 2023). However, Zama Zamas are from all parts of sub-Saharan Africa and South Africa.

The following extracts from Danny, a member of civil society, reflect the misrepresentations around artisanal mining and illegal immigrants:

> "Yes, the thing is the media and the mining industry and government... and I attended meetings of the safety and security sector in Gauteng, where we discussed the topic of Zama Zamas. Where the minister of Safety and Security blamed small-scale micro and survival mining on foreigners... There are people from Lesotho, there are people from Zimbabwe, there are people from Mozambique, there are people from the Eastern Cape, there are people from the Free State, there are people from Gauteng. So, it is a mixture of people."

Indeed, some miners are from neighboring countries like Mozambique and Zimbabwe, but these miners work alongside miners from South Africa (Thornton, 2014). Also, migrant labor, then and now, is not as straightforward as simply labeling miners as illegal immigrants. These movement patterns have a rich and complicated history (Chaps. 3 and 4 discuss this further).

Adding to the complexity of this issue, a 2019 report by Action Aid found that not all Zama Zamas are foreign nationals. Foreign nationals accounted for 29% of the total sample. It was found that most foreign nationals were engaged in gold mining and accounted for over 40% of the total respondents. Lesotho nationals accounted for 54%, while respondents from Mozambique and Zimbabwe accounted for a combined 40%. Interestingly, these results suggest a higher percentage of foreign nationals participating in the gold sector. This might be due to the historical ties of the formal mining industry, which attracted migrant workers to the goldfields of South Africa (see Chap. 3).

In addition, the South African Human Rights Commission and some media coverage have stated that not all Zama Zamas are foreigners or illegal migrants or criminals (SAHRC, 2015; Ledwaba, 2017; Metelerkamp, 2022). The percentages mentioned in the Action Aid report (2019) show the sector's heterogeneous nature and highlight how grossly it is often represented and reported on. Claiming that all miners are illegal immigrants is not only oversimplifying the sector but is entirely erroneous.

Labels and the Creation of Misconceptions

Two unofficial subtypes of artisanal mining exist, namely informal and illegal. However, there is no formal distinction between informal and illegal mining in South Africa, despite scholars' efforts to differentiate between the two (SAHRC, 2015). It has been stated that informal miners mine on customary land with the chief's permission (SAHRC, 2015). Whereas so-called illegal miners trespass on private property.

Since informal mining takes place on customary land under customary laws, it is considered (more) legal, although informal miners are also mining without a mining permit, which means they are operating unlawfully (Wilson, 2018). On the other hand, those Zama Zamas that mine on private land are portrayed in the media as criminals and are, therefore, associated with illicit mining (TimesLive, 2018; Ryan, 2019; Seleka, 2019; Rademeyer, 2022; Mabena, 2023).

Definitions around informal and illegal mining contribute to considering the practice in a country's national legal framework. The Global Initiative Against Transnational Organized Crime (2017) argues that governments wrongly criminalize informal miners by failing to distinguish between informal and illegal mining, thus eliminating the livelihood of a highly vulnerable group. It further states that many countries lack legislation distinguishing between illegal and informal mining. When the vagueness around the definitions clears up, a legal framework could be developed more efficiently. In support of this, the International Institute for Environment and Development (IIED) (2015) remarks that the hazy definitions of legal, illegal, formal, and informal worldwide stop artisanal and small-scale stakeholders from participating in dialogue and collaborative action. Besides, formalization efforts will be problematic without separating legal, criminal, and informal activities, and finding a sustainable way forward will be challenging (IIED, 2015).

A research report by the University of British Columbia (UBC) and the Canadian International Resources and Development Institute (CIRDI) (2019) indicates, to a degree, that distinguishing between informal and illegal mining activities is not straightforward. The report outlines that illegal artisanal and small-scale mining (ASM) ranges from what could be termed informal to criminal. In the former case, the miners are not licensed but want to obtain a license. However, the miners cannot overcome the technical, financial, and administrative barriers to getting a permit. The report states that this form of mining is driven primarily by socioeconomic factors.

On the other hand, the report argues that criminal intent has also been shown to drive participation in ASM, where organized criminal gangs and even international crime syndicates are involved. The report further states that illicit artisanal and small-scale miners engage in illegal surface mining, usually on alluvial diamond and gold deposits or mine dumps, abandoned and disused mine shafts, or active operational shafts on large-scale mines. According to the report, this type of mining is known as Zama Zama mining in South Africa.

Framing Zama Zama mining, as in the latter viewpoint, is too narrow to cover all incidences of artisanal mining in the South African context. It also creates the impression that all artisanal miners mining the mine dumps or in abandoned and disused mine shafts fall on the criminal side of the spectrum. Such thinking is a gross oversimplification and adds to the existing misconceptions around the sector. Most artisanal mining activities in South Africa occur in existing and historical mining communities, mining abandoned mine sites (Action Aid, 2019). This renders the distinction by the UBC and CIRDI report problematic since it deems most miners as criminal and not informal, regardless of intent. By adopting the UBC and CIRDI report's thinking, one risks oversimplifying the artisanal mining sector in South Africa. To illustrate this point, the report by Action Aid (2019) found that artisanal miners, those occurring at abandoned mine sites, engage in artisanal mining because there is a lack of jobs in the formal sector—driven by socioeconomic factors. In those cases, the socioeconomic element serves as a powerful driver for these artisanal miners, not criminal intent, despite them mining mine dumps or abandoned mine shafts.

The issue with such thinking portrayed in the UBC and CIRDI report is that the miners' activities are classified as criminal. Such reports and the fact that Zama Zama mining has also been associated with criminal syndicates intensify misconceptions and create a distorted picture of Zama Zama mining. Ultimately, this serves as the motivation for policing instead of developing the sector. During a conversation with a vice president of sustainability from a mining company, this stance was confirmed:

> "There is no interaction. Because one thing we are clear on is we are all supportive of trying to formalize the issues that are around artisanal and small-scale mining and illegal miners, for that matter, but at the same time, we cannot deal with people perpetrating criminal acts."—Bruce, vice president of sustainable development

The above statement shows that most mining companies perceive artisanal miners as criminals. Ironically, the following quotations by Zama Zamas show that they are engaging in the sector because they don't want to turn to crime. The miners also stated they engage in Zama Zama mining to create a livelihood. Two miners had the following to say:

> "This kind of work is not like we like it. But it is that we don't want to steal from people. So, we are feeding for ourselves."—Sam, Zama Zama

> "For the money. To eat. This life is so difficult."—Tim, Zama Zama

During the discussions, many interviewees stated they did not view the artisanal miners as criminals. In their opinion, the miners were merely trying to make a living:

> "So, we see artisanal mining as one of the ways in which ordinary communities can engage in the mining industry for better livelihoods, so as a livelihood coping strategy."—Themba, civil society

> "Firstly, the artisanal miners, they call themselves Zama Zamas. Zama Zama means somebody who's trying to do some work for them to be able to put bread on the table."—Mike, civil society

During a conversation, I probed Carina, a key informant, about misconceptions in the sector, she commented as follows:

> "It's huge. I mean that's a very difficult question to answer because there are so many layers. Obviously, because it is an illegal and unregulated activity, there is gangsterism; there is murder. But I think there are huge misconceptions. I mean all the people I come across and talk to are just ordinary people. They need to stay alive. And most of them have families. So yes, the misconceptions that's there, to say every Zama Zama is a murderer, is absolutely crazy. They are ordinary people that are caught up in something, which is hugely dangerous with huge poverty."

These insights by the interviewees show that generalized reports must be treated with caution. Skewed statements presented as facts not only maintain misconceptions but also enforce and reinforce the responses from the mining sector and government, which is policing.

It is, therefore, vital to reconsider those artisanal miners classified as criminals by questioning why miners will be mining without a legal permit. According to Jinnah and Tafira (2017), there is no indication that the miners are operating outside the legal framework with the intent to avoid tax and statuary obligations. Instead, the label of illegality results from the absence of effective legislation and regulatory provisions for artisanal mining (Jinnah & Tafira, 2017). Also, The SAHRC (2015) indicated in their report that not all Zama Zamas are involved in criminal syndicates or, if they are, had the intention of becoming part of illegal activities.

Of course, the sector has an element of criminality, and not all Zama Zamas operate with the purest intent. However, the perception of a nearly exclusive focus on the sector's negative impacts forces the industry's potential into the margins of socioeconomic development. In the end, the sector remains underdeveloped and highly marginalized. This predominant focus on the negative impacts, often driven by the government, the formal mining industry, and the press, emphasizes the sector's poor health and working conditions as well as the violence and conflict that pervade the sector.

Conclusion

Zama Zama mining can be considered a symptom of social exclusion perpetuated by labels and misconceptions. Sadly, the Zama Zama mining question is only one of many symptoms of the fractured socioeconomic system where the powerful dominate the powerless. I hope we will begin to identify and amplify the other suppressed voices in our society by casting some light on the Zama Zama matter. This includes women, indigenous groups, informal traders, and others; only by recognizing that a more nuanced version exists (almost always) about nearly anything or anyone can a space be created to voice alternative opinions. In this space, the opportunity to foster change might be birthed.

Navigating the terrain of a book requires a roadmap. This outline summarizes the structure and content of this book. Chapter 2 delves into the migrant labor system that left an indelible mark on South Africa's mining industry. Chapter 3 continues to describe the history of the mining sector, spotlighting the infamous mining hostels. Building upon this foundation, Chaps. 4 and 5 thoroughly explore artisanal mining in South Africa. Chapter 4 describes the distinctive profiles of artisanal miners and what they mine. The chapter also dissects the driving forces behind this sector. Chapter 5 offers an exclusive perspective on women within the artisanal mining sector, recognizing their pivotal but underrepresented contributions within the artisanal mining sector. Chapter 6 sheds light on the challenges and social issues permeating artisanal mining. Following by outlining the challenges and impacts artisanal mining has on the environment and society, Chap. 7 discusses the formalization of the ASM sector, a commonly proposed solution. Governments are not the only role players in developing the artisanal mining sector; for that reason, Chap. 8 looks at the intricate dynamics between ASM miners and their larger-scale counterparts, revealing a complex relationship. Chapter 9, which emphasizes the need for more sustainable practices in artisanal mining and offers pragmatic solutions, concludes this book.

References

Action Aid. (2019). *Regularising informal artisanal mining in South Africa: An evidence-based report*. Action Aid.

Adichie, C. (2009). *The danger of a single story*. TED Talk. Accessed July 20, 2023, from https://www.scribbr.co.uk/referencing/generator/folders/6duzFiHCCe32Q1X0LqIow8/lists/2mjijuSosZmL6UIyW3u7RO/

Babbie, E., & Mouton, J. (2014). *The practice of social research*. Oxford University Press Southern Africa.

Bilton, C. (2007). *Management and creativity: From creative industries to creative management*. http://ci.nii.ac.jp/ncid/BA78183682.

Bourdieu, P. (2019). *Classification struggles: General sociology, Vol. 1 (1981-1982)*. Polity.

Bruhn, J. G., & Rebach, H. M. (1996). *Clinical sociology: An agenda for action*. Plenum Press eBooks. http://ci.nii.ac.jp/ncid/BA2979999X

Chiumbu, S. (2016). Media, race and capital: A decolonial analysis of representation of miners' strikes in South Africa. *African Studies, 75*(3), 417–435. https://doi.org/10.1080/00020184.2016.1193377

Duncan, J. (2014, January 23). South African journalism and the Marikana massacre: A case study of an editorial failure. *The Political Economy of Communication, 1*(2), 65–88. http://www.polecom.org/index.php/polecom/article/view/22/198

Enviro Editor. (2018). *The wake of South Africa's abandoned gold mines*. Accessed January 15, 2019, from https://www.environment.co.za/mining/wake-south-africas-abandoned-gold-mines.html

Foucault, M. (1972). *Archaeology of knowledge and the discourse on language*. Tavistock Publications Limited. https://monoskop.org/images/9/90/Foucault_Michel_Archaeology_of_Knowledge.pdf

Foucault, M. (1981). The order of things. In R. Young (Ed.), *Untying the text: A post-structuralist reader* (1st ed., pp. 51–79). Routledge & Kegan Paul. http://www.kit.ntnu.no/sites/www.kit.ntnu.no/files/Foucault_The%20Order%20of%20Discourse.pdf

Fritz, J. M. (1989). The history of clinical sociology. *The Development of Clinical and Applied Sociology, 7*(1) http://digitalcommons.wayne.edu/socprac/vol7/iss1/11

Fritz, J. M. (2008). Basics: From concepts to models. In J. M. Fritz (Ed.), *International clinical sociology* (1st ed., pp. 7–18). Springer.

Fritz, J. M. (2013). Sociological practice and public sociology. In D. L. Brunsma, K. E. Iyall Smith, & B. K. Gran (Eds.), *Handbook of sociology and human rights* (1st ed., pp. 394–401). Routledge. https://doi.org/10.4324/9781315634227

Germeten, S. (2013). Personal narratives in life history research. *Scandinavian Journal of Educational Research, 57*(6), 612–624. https://doi.org/10.1080/00313831.2013.838998

Giddens. (2001). *Sociology* (4th ed.). Polity.

Global Initiative against Transnational Organized Crime. (2017). Global Initiative against Transnational Organized Crime. https://globalinitiative.net/wp-content/uploads/2017/05/global-initiative-annual-report-2016.pdf

Hall. (2006). Narrative as vital methodology in clinical sociology. *Journal of Applied Sociology, 23*(1), 53–67. http://www.jstor.org/stable/43736124

Held, J. M. (2016). Introduction: On writing popular philosophy. In J. M. Held (Ed.), *Stephen King and philosophy* (pp. 1–13). Rowman & Littlefield.

Honneth, A. (1991). *The critique of power: Reflective stages in a critical social theory*. The MIT Press.

Honneth, A. (1996). *The struggle for recognition: The moral grammar of social conflicts* (1st ed.). MIT Press.

Honneth, A. (2004). Organized self-realization: Some paradoxes of individualization. *European Journal of Social Theory, 7*(4), 463–478. https://doi.org/10.1177/1368431004046703

Honneth, A. (2014). *Freedom's right: The social foundations of democratic life* (1st ed.). Polity.

International Institute for Environment and Development (IIED). (2015). *ASM – LSM – Government relations*. Accessed February 22, 2017, from https://iied.org/sites/default/files/pdfs/migrate/G03938.pdf

Jinnah, Z., & Tafira, K. (2017, March 31). *An ethnography of informal artisanal gold mining in Johannesburg*. SEAM. Accessed June 7, 2017, from https://seampartnership.files.wordpress.com/2017/05/work-and-well-being-on-the-margins1.pdf

Johnson, C., & Forsyth, T. (2002). In the eyes of the state: Negotiating a "rights-based approach" to forest conservation in Thailand. *World Development, 30*(9), 1591–1605. https://doi.org/10.1016/s0305-750x(02)00057-8

Lewaba, L. (2017, August 9). Zama-zamas take their chances rather than starve. *Mail & Guardian*. Accessed November 2, 2024, from https://mg.co.za/article/2017-08-09-zama-zamas-take-their-chances-rather-than-starve/?form=MG0AV3

Mabena, S. (2023). Zama zamas spread their roots as illegal mining moves across provinces. *The Citizen*. Accessed October 30, 2024, from https://www.citizen.co.za/news/south-africa/zama-zamas-spread-their-roots-as-illegal-mining-moves-across-provinces/

Metelerkamp, T. (2022, February 2). The zama zama: Informal mining 'unlike anything else in the world. *Daily Maverick*. Accessed November 2, 2024, from https://www.dailymaverick.co.za/article/2022-02-02-the-zama-zama-informal-mining-unlike-anything-else-in-the-world/?form=MG0AV3

Minerals Council South Africa. (2023). Illegal Mining. https://www.mineralscouncil.org.za/work/illegal-mining

Rademeyer, J. (2022, August 7). Zama zamas just one part of a multibillion-rand organised crime economy that threatens SA. *Daily Maverick*. Accessed October 30, 2024, from https://www.dailymaverick.co.za/article/2022-08-07-zama-zamas-just-one-part-of-a-multibillion-rand-organised-crime-economy-that-threatens-sa/

Rhéaume, J. (2014). Research for the community. In J. M. Fritz & J. Rhéaume (Eds.), *Community intervention: Clinical sociology perspectives* (pp. 31–43). Springer.

Rodny-Gumede, Y. (2015). *Marikana massacre: How South African journalism failed the test*. Press Release. https://theconversation.com/marikana-massacre-how-south-african-journalism-failed-the-test-51130.

Ryan, C. (2019, March 6). Cable theft and illegal mining threaten Joburg's infrastructure. *Moneyweb*. https://www.moneyweb.co.za/mineweb/cable-theft-and-illegal-mining-threaten-joburgs-infrastructure/.

Seleka, N. (2019, January 9). 6 Zama Zamas Arrested During Shootout Between Rival Gangs. *News24*. Accessed March 23, 2021, from 6 zama zamas arrested during shootout between rival gangs | News24

Snyder, C. (1984). Analyzing classifications: Foucault for advanced writing. *College Composition and Communication, 35*(2), 209. https://doi.org/10.2307/358097

South African Human Rights Commission (SAHRC). (2015). *Report of the SAHRC investigative hearing issues and challenges in relation to unregulated artisanal underground and surface mining activities in South Africa*. Accessed November 12, 2015, from https://www.sahrc.org.za/home/21/files/Unregulated%20Artisanal%20Underground%20and%20Surface%20Mining%20Activities%20electronic%20version.pdf

The University of British Columbia and Canadian International Resources and Development Institute (CIRDI). (2019). *Artisanal and small-scale mining: mapping the South African ecosystem*. Accessed November 19, 2021, from https://www.delvedatabase.org/resources/artisanal-and-small-scale-mining-mapping-the-south-african-ecosystem

Thornton, R. J. (2014). Zamazama, "illegal" artisanal miners, misrepresented by the South African Press and Government. *The Extractive Industries and Society, 1*(2), 127–129. https://doi.org/10.1016/j.exis.2014.06.003

Timeslive. (2018, May 9). Zama zama gold processing hub shut down. *TimesLIVE*. https://www.timeslive.co.za/news/south-africa/2018-05-09-zama-zama-gold-processing-hub-shut-down/

Uys, T. (2021). The emergence of clinical sociology in South Africa. In J. M. Fritz (Ed.), *International clinical sociology* (2nd ed., pp. 109–127). Springer.

Wilson, L.-A. (2018, January 1). *Unshackling South African artisanal miners: Considering Burkina Faso's legislative provisions as a guideline for legalisation and regulation*. MA thesis, University of Cape Town. https://open.uct.ac.za/handle/11427/29638.

Webster, E., Lambert, R., & Bezuidenhout, A. (2008). *Grounding globalisation: Labour in the age of insecurity*. Blackwell.

Chapter 2
Historical Laws and Practices That Influenced South Africa's Mining Sector

History, as we all know, does not repeat itself. Every now and then, however, it bears rereading with a sharp eye, looking for underlying continuities—be they ideological, political, economic or social.

—van Onselen (2001)

Introduction

Many factors shape and shift the dynamics of a country's artisanal mining sector. No artisanal mining operation in the same country is identical, let alone cross-country operations. There will be overlapping characteristics and cases where lessons can be drawn from artisanal mining sites; however, each country differs in many ways, such as economic, social, and political challenges and opportunities. Apart from these meta factors, a government's stance toward the artisanal mining sector is another element that adds to the dynamic of a country's artisanal mining sector. History is another factor that influences the subtleties of a country's artisanal mining sector, sometimes one that seems to be disregarded.

Historical factors play a role in the trajectory path of development. It is by no means an exaggeration to say that history does matter. Considering the past and the importance of narratives, Wilson (2001) draws our attention to distinguishing between two dominant narratives around South Africa's economic development.

From one point of view, the South African economy's expansion throughout the first century of its industrialization, ignited by the discovery of gold in 1886, was highly successful. There is, however, another side to the story. What is so sad about South Africa's progress is the extent to which the exact methods that produced economic wealth also generated poverty, exclusion, and unemployment patterns (Wilson, 2001).

Some scholars have distinguished between two central arguments about the mining industry in South Africa's socioeconomic trajectory: voluntarists and revisionists (Delius, 2017; Wilson, 1972). The voluntarists ascribed the occurrence of migrant labor in South Africa to market forces at work in the context of underdeveloped local economies in tribal areas. The revisionists emphasized the critical role of conquest, land dispossession, increasing taxes aided by Draconian pass laws, centralized recruiting, compounds, and divide-and-rule tactics in addressing the causes

V. Bester, *The Untold Story of Zama Zama Miners in South Africa*, Clinical Sociology: Research and Practice, https://doi.org/10.1007/978-3-031-82783-9_2

of the system's expansion (Delius, 2017). A second wave of revisionists, led by Harold Wolpe and Colin Bundy, argued that the need for cheap labor fueled the development of segregation and, ultimately, the apartheid regime (1948–1994) (Vosloo, 2020).

In the same vein, Adler et al. (2007) note that the early gold mining industry has been described as "*... an extractive industry with little consideration given to possibly adverse long-term effects.*"

Before 1994, the mining industry personified some of the most damaging social practices (Siyobi, 2015). Indeed, the system had little or no room for embedding a social conscience for the present, at that point, and as we will see, nor for the future. Vosloo (2020) rightly remarks that dismantling the migrant labor system and undoing the harm it caused, like so many other political and social systems, often necessitates even more tenacious efforts over time. South Africans who are justly engaged in restitution for past economic and social injustices may need to adopt a wider perspective (van Onselen, 2019).

Hence, this chapter takes a retrospective approach, offering a broader contextualization of South Africa's artisanal mining sector. It achieves this by outlining the historical and social injustices stemming from past laws in the country, which continue to echo through the mining landscape. Legislation was crafted to disempower Black people and subsequently had profound repercussions on that era's Black mining labor force. It is vital to acknowledge the historical background because the one-dimensional view of the current Zama Zama issue as insular, a nuisance and menace, and with no deep-seated mining history is mistaken. It creates the narrow, one-sided perspective Wilson (2001) refers to and ignores the other side, such as past and present socioeconomic injustices and how they were shaped. The first part of this chapter briefly looks at laws that either facilitated or upheld the migrant labor system. Following this short overview, the second part of this chapter delves deeper into the migrant labor system itself.

Native Labour Regulation Act (1911)

The Native Labour Regulation Act of 1911, a pivotal piece of legislation in the history of South Africa, holds a profound legacy that continues to echo through the country's social and economic landscape. This Act, was emblematic of the broader system of racial segregation and exploitation that characterized South Africa during the early twentieth century; it also had far-reaching consequences for the rights and mobility of Black South Africans, particularly in the context of labor and employment. It was a legislative cornerstone of the broader system of apartheid that would later be formalized in the mid-twentieth century.

In addition, this Act was enacted to help maintain an adequate supply of migrant labor for the gold mines and centralize government control over African laborers in industrial areas. Furthermore, the Act gave the Director of Native Labor the authority to issue licenses to labor recruiters and employers who hired workers (Duncan,

1993). While the Native Labour Regulation Act of 1911 had not outlawed industrial action entirely, it curbed it enough to ensure that most strikes were largely unsuccessful.

The Act was primarily aimed at mineworkers and specified that a breach of a labor contract could be a criminal offense. This prohibited mine workers from leaving the mines, which 15% did before 1911. However, under the Act, a worker who had abandoned his work at the mines could be traced back to his home and jailed. After jail time, the worker could be returned to the mines to finish his contract (Van Dyk, 2018). Another law that served the migrant labor system well was the taxes imposed on communities from which the mining industry drew its workforce.

Special Taxes

The rise of Johannesburg as a mining hub in 1886 gave way to a new and powerful interest group, a speedily evolving body of mine owners seeking labor (Ndlovu, 2017). To lure more men to the mines from the countryside, the government then introduced a poll tax and a hut tax that demanded a fee Black communities had to pay for each hut in their community (CJPME Foundation, 2014).

Furthermore, colonial administrations relied on the levy fees paid by the Chamber of Mines' recruitment arm, known as the *Witwatersrand Native Labour Association* (WNLA), for each recruit. The colonial administration also relied on the mines to help them collect overdue hut tax (McCulloch, 2014). There has been debate over whether the hut tax was intentionally designed to induce a flow of labor to South Africa's mining industries (Massey, 1978). An analysis of the historical records from this period shows that the aims of introducing taxation were multi-layered (Ndlovu, 2017).

Regardless of the intention behind the hut tax, the officials collecting the hut tax knew that introducing such a tax would necessitate a migrant labor movement (Massey, 1978). The hut tax, for example, drove African men into wage labor and was one of the forerunners of the coercive labor control procedures that characterized the apartheid system later. Gwaindepi and Siebrits (2020) write that the government indirectly incentivized men to look for work within the migrant labor system by introducing various forms of colonial taxes, such as the hut and poll taxes.

A revision of the hut tax in 1895 was aimed at compelling African men to migrate to South Africa for mine labor. Under the new rule, unmarried men over 21 years old, or those with one wife, were required to pay 10 shillings. In addition, a poll tax of 40 shillings was levied on males aged 21 years and older.

Other legislation ultimately supplemented hut taxes to control the supply of African labor to the mines in the Cape Colony and later in the Union of South Africa (Gwaindepi & Siebrits, 2020). During this time, the availability of low-cost African labor was critical to secure profitability for the mines.

Another law that supported the flow of migrant labor was the introduction of passes. Once again, the role mining companies played in this matter is conspicuous.

The Introduction of Passes

South Africa's pass laws were an internal passport system restricting Black people's travel by forcing them to carry passbooks while leaving their homelands or designated territories. Primarily associated with the apartheid era, the roots of the pass laws go back to the earlier colonial times and slave labor (Institute for Justice and Reconciliation 2006).

The first pass rules were enacted in 1760 and applied to enslaved Africans living in the Cape Colony. Then, in 1809, the Governor of the Cape Colony issued a law requiring all Africans to live in a single location. They had to have a pass if they relocated. By 1827, all Africans arriving from outside the Cape Colony required a pass. These regulations were enacted to regulate people's movement into the Cape Colony.

This strategy was created to provide a steady supply of migrant labor while segregating the population. This regulation was utilized during the apartheid era to limit the number of African residents in urban areas. Prior to the 1950s, this legislation solely applied to African men. Furthermore, the mining industry's misleading and coercive recruitment campaigns demanded additional efforts to ensure that migrant workers met their duties. They backed the Transvaal Government's reintroduction of the Pass and Master and Servant Laws. Restoring these laws gave them the broad authority to exercise complete control over labor (Vosloo, 2020).

Although rules requiring Black people to have passes had been adopted before the discovery of diamonds in Kimberley, the requirement was more regularly enforced after the diamond and gold booms, which restricted Black people from moving into cities (Warnsley, 1996; CJPME, 2014).

Attempts to extent the pass laws to women in the 1950s were met with strong opposition, including the iconic 1956 Women's March. Although enforcement varied during that decade, by the 1960s, pass laws were being implemented more strictly for women. In 1986, pass laws were effectively removed for both men and women (Ndlovu, 2017).

The Notorious Migrant Labor System

Before formally establishing the migrant labor system, a trade network emerged in Southern Africa around 1850 due to new economic hubs and increased demands for labor (Vosloo, 2020). This shift was driven by the expansion of the commercial wool production industry and the burgeoning sugar cane sector in what is now

KwaZulu-Natal, South Africa. These changes ultimately paved the way for establishing the migrant labor system, which colonial and postcolonial politicians and capitalists subsequently co-opted to serve their interests. Delius (2017) highlights a long-standing historical pattern of individuals seeking employment and income outside their native regions, a practice that continues in present-day South Africa.

According to Dlamini (2014), the South African migrant labor system has existed since the discovery of diamonds in Kimberley in 1868. It has been characterized as a historical structure created by capitalist, colonial, and pre-apartheid authorities to address the pressing need for cheap labor in the then-White South African-owned mines and towns. This system forced Black people, primarily boys, and adult men, to migrate and work away from their home areas. However, migrant workers could not move with their families or stay permanently in the working areas. This caused a system of alternating migrant labor (Vosloo, 2020).

The mining companies proclaimed that South Africans would not work for the mining industries' wages. Therefore, foreign labor was preferred. Besides, the mining companies contended that foreign workers comprised a reservoir of skills, experience, and work discipline that the mining industry could not do without. Later, almost 80% of the migrant labor force came from outside South Africa in the 1970s (see Crush et al., 2005).

Before we explore the details of the migrant system, it is necessary to understand why the system was used. Apart from the reason mentioned above, perhaps more sinister, was another reason behind this well-oiled migratory system that engulfed people from afield. The motivations for migrant labor are best outlined in the following statement, as explained by the then Chamber of Mines (today known as the Council of Minerals) (Wolpe, 1972):

> It is clearly to the advantage of the mines that native laborers should be encouraged to return to their homes after the completion of the ordinary period of service. The maintenance of the system under which the mines are able to obtain unskilled labor at a rate less than ordinarily paid in an industry depends upon this, for otherwise, the subsidiary means of subsistence would disappear, and the laborer would tend to become a permanent resident upon the Witwatersrand, with increased requirements…

The drive behind the migrant labor system had more to do with revenue than skills, as contended by the mining industry of that time. The system was characterized by a bottomless hunger for profit and little regard for workers who generated income for investors and mining magnates.

Mozambique has served as a consistent, fertile recruiting ground for the mines since the discovery of diamonds in Kimberley in 1868 (Dlamini, 2014). Less than 10 years later, by 1874, there were about 10,000 Black migrant workers on the Kimberley diamond fields (Wilson, 1972 as cited in Vosloo, 2020). Between 1886 and 1936, most men working in the mines came from Mozambique (Dlamini, 2014).

Furthermore, by 1899, 13 years after discovering gold on the Witwatersrand, the gold mines employed 100,000 Black migrant workers (Wilson, 1972). The increase in the number of workers that came to the Witwatersrand during the late 1800s is astonishing. It has been reported that "… *between 1890 and 1899 the total number of Africans employed on the Witwatersrand gold mines rose from 14,000 to 97,000*"

(Wentzel & Tlabela, 2005). These are small numbers compared to what will come when the mining industry of South Africa boomed.

As mentioned, Mozambique played a pivotal role as a primary source of labor for South African mining companies. Charles van Onselen (2019) underscores the sheer magnitude of Mozambican laborers employed in the gold mines in his book *The Night Trains.* van Onselen rightly emphasizes that all South Africans, especially those who have historically owned and still possess coal and gold mining enterprises, must acknowledge that a significant portion of the nation's historical prosperity, wealth, and modern infrastructure was built upon the toil of Black laborers extracted from colonial Mozambique.

Part of the rationale behind this influx of Mozambican labor into South Africa's migrant labor system lies in the stark choice these men faced: either submit to forced labor, as sanctioned by the Portuguese colonial regime at that time or opt for migrant labor. Despite the arduous and perilous conditions in the mines, the latter was still perceived as a preferable alternative to forced labor.

Since the start of industrialized mining, the mines have employed many migrant workers drawn from rural communities and neighboring countries. So ample was the supply from neighboring countries that between 1911 and 1990, approximately 50% of recruits were from South Africa's neighboring countries (McCulloch, 2014). Moreover, it has been estimated that around 1903, the number of Black workers working in the gold mines was around 77,000. By the late 1920s, the number stood at 200,000. By the 1930s, the number rose to over 300,000. It more or less stayed at this figure well into the 1950s (van Onselen, 2019). The workforce peaked at over half a million in the late 1970s and declined sharply afterward. The historic workforce consisted of unionized White labor in supervisory roles and Black migrants who did the manual work (McCulloch, 2014).

A civil society worker, mentioned the matter of migrant labor during a discussion by explaining how the South African mining industry has historically depended on it since its inception in 1886.

> "Since 1886 when industrial mining started in Gauteng. The mining industry has been reliant on migrant labor. So generations of migrant workers have come from Zimbabwe, come from Zambia, come from Malawi from, Mozambique from Lesotho and so on into South Africa to come and work in the mines. And it has become almost a cultural thing in many of these communities where the local chief was actually the recruiting agent for the labor recruiting agency. And it was almost seen as a ritual to manhood."—Danny, civil society

The reference to the cultural dimension underscores how migration became deeply embedded in the fabric of communities. The local chief's role as a recruiting agent signifies the institutionalization of this labor system within traditional structures. This practice was not merely economic but also a significant cultural practice, marking the transition to manhood for many young men.

Table 2.1 shows the countries and corresponding numbers from 1920 to 1990, from which South Africa drew its migrant labor.

According to Table 2.1, the South African mining industry recruited migrants from nearly every other country. Following independence, most nations reconsidered the issue of labor migration to other countries. As per the table, Mozambique

Table 2.1 South African labor-sending countries (1920–1990) (Crush et al., 2005) (numbers revised by author, 2024)

Year	Angola	Botswana	Lesotho	Malawi	Mozambique	Swaziland	Tanzania	Zambia	Zimbabwe	Other	Total
1920	0	2112	10,439	354	77,921	3449	0	12	179	5484	99,950
1925	0	2547	14,256	136	73,210	3999	0	4	68	14	94,234
1930	0	3151	22,306	0	77,828	4345	183	0	44	5	107,862
1935	0	7505	34,788	49	62,576	6865	109	570	27	9	112,498
1940	698	14,427	52,044	8037	74,693	7152	0	2725	8112	70	167,958
1945	8711	10,102	36,414	4973	78,588	5688	1461	27	8301	4732	158,997
1950	9767	12,390	34,467	7831	86,246	6619	5495	3102	2073	4826	172,816
1955	8801	14,195	36,332	12407	99,449	6682	8758	3849	162	2299	192,934
1960	12,364	21,404	48,842	21934	101,733	6623	14,025	5292	747	844	233,808
1965	11,169	23,630	54,819	38580	89,191	5580	404	5898	653	2686	232,610
1970	4125	20,461	63,988	78492	93,203	6269	0	0	3	972	267,513
1975	3431	20,291	78,114	27904	97,216	8391	0	0	2485	12	237,844
1980	5	17,763	96,309	13569	39,539	8090	0	0	5770	1404	182,449
1985	–	18,079	97,639	16849	50,126	12,365	0	0	0	4	195,062
1990	0	15,720	108,780	72	50,104	17,816	0	0	2	0	192,494
Total	59,071	203,777	789,537	231,187	1,151,623	109,933	30,435	21,479	28,626	23,361	2,649,029

served as South Africa's major labor-sending country, with over 1 million workers from 1920 till 1990. The second-largest portion of migrant workers were from Lesotho, with nearly 800,000 workers for the same period. Some countries, including Tanzania, Malawi, and Zambia, attempted to prevent migrant labor (Crush et al., 2005).

After independence, Tanzania and Zambia withdrew their workers from the South African mines. Malawi recalled all of its employees in 1972. But, following a renegotiated deal in the late 1970s, around 20% of the miners returned to the mines. Employment persisted at this level until 1987 when the Malawian government withdrew the workers in response to a disagreement with the apartheid administration over human immunodeficiency virus (HIV) testing.

In the 1980s, the supplier countries established the Southern African Labour Commission (SALC) to develop a uniform policy on labor migration to South Africa. However, its efforts to adopt a phased disengagement program were largely unsuccessful. Countries like Lesotho and Mozambique, could not fully end contract labor migration (Crush et al., 2005).

To sum up, during pre-apartheid and apartheid, rural areas and South Africa's neighboring countries served as sources to meet the demands of the booming mining industry. Labor was also extensively drawn from the then homelands in South Africa and, as we have seen, South Africa's neighboring countries. It has been noted that more recently (post-apartheid), some of these areas have become South Africa's new mining frontiers, with increased mining activity in many former homeland areas (Mnwana & Bowman, 2022). Mining activities in these include coal and titanium mining in the former Zululand (now Kwa-Zulu Natal), titanium in the former Transkei (now Eastern Cape), and—Platinum Group Metals (PGM) mining in the former Lebowa and Bophuthatswana homelands (now North West Province and Limpopo) (Mnwana & Bowman, 2022).

Despite the current mining activities in those regions, the socioeconomic situation in these former homeland districts is dire, with high unemployment and poverty levels. Scully and Britwum (2019) assert that the labor reserve areas of the past and present are still relevant today, not because they serve as a labor basket but because their functional utility is disappearing. Mnwana and Bowman (2022) explain that this is mainly due to the prevalence of unemployment in the face of increased capital intensity and a significant decline in labor demands in the historically dominant industries, particularly in the mining sector. Rural residents have now been transformed into quintessential examples of *"surplus populations"* meaning they are more than the requirements of capitalist accumulation (Scully & Britwum, 2019). They were regarded as crucial in supplying labor, notably *"cheap labor"* for the urban industrial and extractive sectors and capitalist export-oriented agriculture, but have subsequently lost their functionality (Burawoy, 1976; Wolpe, 1972).

Regarding the Zama Zama and (illegal) immigrant issue in South Africa, it has been documented that a high percentage of foreign nationals mining gold as artisanal miners could be connected to the historic migrant labor system of the gold mining companies (Action Aid, 2019). Moreover, it has been aptly noted that Zama Zama mining in South Africa is influenced by a complex interplay of factors, deeply rooted in history, with the migrant labor system playing a significant role in its development (Martin, 2019).

Also, there is a generational legacy within the mining industry. Mining is often passed down through families, with young men identifying closely with their fathers' occupations. During a discussion with a key informant, it was mentioned that many Zama Zamas are from a mining family. She said the following:

> "…you come across young men who will tell you that my father was a miner."—Tessa, key informant

Zama Zama mining cannot be separated from the historical and socioeconomic landscape that has shaped it. Furthermore, the generational legacy within mining families perpetuates this tradition, embedding mining deeply into the cultural fabric.

Conclusion

The story of South African mining is not solely one of grandeur and prosperity. It is also a narrative stained by exploitation, social inequalities, and the systemic marginalization of workers, mainly Black South Africans and migrant workers from other parts of Africa. The early mining ventures, driven by the desire for wealth and economic prosperity, laid the groundwork for establishing formal mining industries. These industries, often characterized by capital-intensive and mechanized operations, significantly shaped the nation's economy. The historical legacies of racial segregation, pre-apartheid era policies, and the migrant labor system cast long shadows over the mining sector that created deep-rooted socioeconomic disparities. While masquerading as a tool for economic expansion, the migrant labor system created the perfect breeding ground for South Africa's apartheid system. Cooke (2007) describes the system's aims as *"perhaps the most destructive social engineering of the country's history."*

South Africa's formal mining sector was built on migrant labor, and immigrants were lured to the country when it served the state and formal mining industry. A complex mix of historical factors shapes Zama Zama mining in South Africa, with the migrant labor system being a key influence in its evolution. It can be argued that when immigrants are part of the artisanal mining sector, the past migrant labor system should be considered within such a context. Also, should there be illegal immigrants in the artisanal mining sector, it is up to the government to address the matter in a humanitarian way. The sector cannot be mindlessly criminalized because it is assumed and thoughtlessly accepted that every Zama Zama is an illegal immigrant. Policies must be implemented to address this in cases where illegal immigrants are prevalent.

The legacy of mining in South Africa is multifaceted, and any attempt to address the challenges and opportunities presented by artisanal mining must be grounded in a deep understanding of this historical context. Chapter 3 examines the historical power dynamics within the formal mining industry, acknowledging the centrality of control in its actions. The chapter offers a detailed perspective by exploring some of the past actions by mining companies and the exploitative nature of the mining industry's historical track.

References

Action Aid. (2019). *Regularising informal artisanal mining in South Africa: An evidence-based report*. Action Aid.

Adler, R., Claassen, M., Godfrey, L., & Turton, A. (2007). Water, mining, and waste: An historical and economic perspective on conflict management in South Africa. *The Economics of Peace and Security Journal, 2*(2), 10.15355/epsj.2.2.33.

Burawoy, M. (1976). The functions and reproduction of migrant labor: Comparative material from Southern Africa and the United States. *American Journal of Sociology, 81*(5), 1050–1087.

CJPME Foundation. (2014). *Roots of apartheid: South Africa's mining industry*. Accessed December 20, 2021, from Roots of Apartheid: South Africa's Mining Industry - The CJPME Foundation

Cooke, J. (2007). The form of the migrant labour hostel. *Architecture South Africa, July/August*, 64–69.

Crush, J., Williams, V., & Peberdy, S. (2005). *Migration in South Africa*. Accessed December 21, from https://www.iom.int/sites/g/files/tmzbdl486/files/2018-07/RS7.pdf

Delius, P. (2017). *The History of migrant labor in South Africa (1800–2014)*. Oxford Research Encyclopedia of African History. https://doi.org/10.1093/acrefore/9780190277734.013.93.

Dlamini, J. (2014). Stray boys: The Kruger National Park and migrant labour. In P. Delius, L. Phillips, & F. Rankin-Smith (Eds.), *A long way home: Migrant worker worlds 1800–2014* (pp. 132–143). Wits University Press. https://doi.org/10.18772/22014077670.13

Duncan, D. (1993). The regulation of working conditions for Africans in South Africa, 1918–1948. *African Studies, 52*(1), 1–25. https://doi.org/10.1080/00020189308707764

Gwaindepi, A., & Siebrits, K. (2020). 'Hit your man where you can': Taxation strategies in the face of resistance at the British Cape Colony, c.1820 to 1910. *Economic History of Developing Regions, 35*(3), 171–194. https://doi.org/10.1080/20780389.2020.1791699

Martin, A. (2019). Uncovered: The dark world of the Zama Zamas. https://enactafrica.org/research/policy-briefs/uncovered-the-dark-world-of-the-zama-zamas.

Massey. (1978). *A case of colonial collaboration: The Hut tax and migrant labour on JSTOR*. Botswana Notes and Records. https://www.jstor.org/stable/40979541

McCulloch, J. (2014). The illusion of safety: Migrant labour and occupational disease on South Africa's gold mines. In P. Delius, L. Phillips, & F. Rankin-Smith (Eds.), *A long way home: Migrant worker worlds 1800–2014* (pp. 106–115). Wits University Press.

Mnwana, S., & Bowman, A. S. (2022). Land, conflict and radical distributive claims in South Africa's rural mining frontier. *The Extractive Industries and Society, 11*, 100972. https://doi.org/10.1016/j.exis.2021.100972

Ndlovu, T. (2017). *Fiscal histories of sub-Saharan Africa: The case of South Africa*. Accessed December 24, 2022, from Fiscal-Histories-240817-1.compressed.pdf (pari2.wpengine-powered.com)

Scully, B., & Britwum, A. O. (2019). Labour reserves and surplus populations: Northern Ghana and the Eastern Cape of South Africa. *Journal of Agrarian Change, 19*(3), 407–426. https://doi.org/10.1111/joac.12309

Siyobi. (2015, September 2). *Corporate Social Responsibility in South Africa's mining industry: An assessment – SAIIA*. SAIIA. https://saiia.org.za/wp-content/uploads/2015/09/saia_spb_142_siyobi_20150929.pdf.

Van Dyk, M. C. (2018). *Race, resistance and the general tax of 1925: A historical overview of the interpretation and implementation of South African poll tax*. University of South Africa. http://hdl.handle.net/10500/25336

van Onselen, C. (2001). *New Babylon New Nineveh*. Jonathan Ball Publishers.

van Onselen, C. (2019). *The night trains*. Jonathan Ball Publishers.

Vosloo, C. (2020). Extreme apartheid: the South African system of migrant labour and its hostels. *Image & Text, 34*, 10.17159/2617-3255/2020/n34a1.

Warnsley, J. R. (1996). *Reflections on apartheid in South Africa: Perspectives and an outlook for the future. A curriculum unit*. Fulbright-Hays Summer Seminar Abroad 1996 (South Africa). Accessed December 24, 2022, from https://files.eric.ed.gov/fulltext/ED415168.pdf.

Wentzel, M., & Tlabela, K. (2005). Historical background to South African migration. In P. Kok, D. Gelderblom, J. Oucho, & J. Van Zyl (Eds.), *Migration in South and Southern Africa*. HSRC Press.

Wilson, F. (1972). *Migrant labour: Report to the South African Council of Churches*. Ravan Pr of South Africa.

Wilson, F. (2001). Minerals and migrants: How the mining industry has shaped South Africa. *Daedalus, 130*(1), 99–121. http://www.jstor.org/stable/20027681

Wolpe, H. (1972). Capitalism and cheap labour-power in South Africa: From segregation to apartheid[1]. *Economy and Society, 1*(4), 425–456. https://doi.org/10.1080/03085147200000023

Chapter 3
Unsettled Lives and South Africa's Migrant Hostels

> *My work with the poor and the incarcerated has persuaded me that the opposite of poverty is not wealth; the opposite of poverty is justice.*
>
> —Stevenson (2014)

Introduction

Historically, specific laws and practices in South Africa were developed to serve that era's mining industry economically. These practices, deeply rooted in apartheid, created a robust capital base for the South African mining industry, as Stewart, Bezuidenhout and Bischoff (2019) have similarly highlighted. These laws and practices underscore the industry's tumultuous past and reveal a disturbing legacy of social injustices, notably exemplified in the construction and use of the historic mining hostels.

The mining sector in South Africa has always wielded significant political influence, predominantly controlled by a handful of multinational corporations. For a considerable period, the mines held the dual role of South Africa's largest employer and key source of foreign exchange, wielding considerable authority in shaping the country's economic landscape.

This chapter seeks to unveil the profoundly ingrained power structures within the mining industry by looking at specific historical practices prevalent at the time. The first section of this chapter explores the Mines and Works Act of 1911. This seminal legislation, enacted in the early twentieth century, laid the cornerstone for a racially segregated labor system that endured for decades, significantly shaping the mining industry's trajectory. Subsequently, the chapter shifts its focus to the mining hostels in South Africa, acting as symbols of the intricate interplay between labor, race, power, and the extractive industries in the nation's history. These hostels are tangible reminders of the complex socioeconomic and racial dynamics that have defined South Africa's mining landscape. Following this section, an overview of the grim health and safety circumstances of the early miners is presented. In response to these historic atrocities, the chapter's final section focuses on post-apartheid laws implemented to address past injustices. The chapter concludes by discussing the current socio-legal position of artisanal miners in South Africa.

V. Bester, *The Untold Story of Zama Zama Miners in South Africa*, Clinical Sociology: Research and Practice, https://doi.org/10.1007/978-3-031-82783-9_3

Mines and Works Act (1911) and Its Amendments (1926)

The Mines and Works Act of 1911 was rooted in the British colonial legacy and the economic interests of the mining industry. Scholars such as Wilson (1972) have extensively documented the historical development of labor relations in South Africa, highlighting how the mining industry became a focal point of racial discrimination and economic exploitation.

The Mines and Works Act, passed in 1911, is also known as the *Color Bar* or *Color Bar Act*. The term color bar usually refers to: "*a group of labor practices, informal trade union practices, government regulations, and legislation, all of which were developed over time to prevent Blacks from competing for certain categories of jobs monopolized by Whites*" (Crush et al., 1991).

This legislation fundamentally reshaped the South African mining labor force along racial lines. It established distinct categories of work for different racial groups, with White workers being designated for skilled and supervisory roles, while Black workers were relegated to unskilled and often hazardous positions. As Meredith (2007) explained that Black workers were subjected to legislation designed to lower them to a strictly subordinate role and exploit their labor potential. This was achieved by stipulating that certificates for skilled trades in the mine could only be issued to White workers for specific occupations (CJPME Foundation, 2014).

The Act aimed to prevent Black workers from competing with White workers for better-paid jobs (CJPME Foundation, 2014). The Act also introduced stringent restrictions on the mobility of Black laborers. Under its provisions, Black workers were required to carry passes, akin to identification documents, that specified their employment status and location (see Chap. 2).

In 1926, the Mines and Works Act was revised. The then prime minister J.B.M. Hertzog's White working-class backing was guaranteed by the Wage Act of 1925 and the Mines and Works Act of 1926. The Wage Act protected low-wage, unskilled White workers, and the Mines and Works Act shielded skilled and semi-skilled White workers against Black worker competition. At that time, an opposition member of parliament rightly pointed out that the government denied Black people access to specific jobs while raising their financial burden by implementing the 1925 poll tax (Van Dyk, 2018).

The Act had profound economic implications. While it provided a steady and cheap source of labor for the mining industry, it perpetuated racial inequalities and limited opportunities for Black workers to advance in their careers. This exploitative labor system contributed to the vast disparities in income and living standards between racial groups in South Africa.

As the demand for cheap labor grew, the need for housing developed. The migrant labor hostels emerged and became essential accommodations for workers who journeyed from rural areas to urban mining centers. These hostels embody the labor segregation and social inequalities institutionalized by the Mines and Works Act, underscoring the complex link between this legislation and the living conditions of the labor force it governed. Although not in use anymore, the mining hostels still serve as a stark reminder of the grim history of South Africa's formal mining industry.

The Historical Mining Compounds

The mining compound system in South Africa was introduced in the late nineteenth century to regulate and oversee the labor force in the rapidly expanding diamond and gold mining sectors. This system aimed to maintain a consistent supply of affordable, dependable labor while preventing activities like diamond smuggling. The Kimberly hostels for diamond miners served as a testing ground for the Witwatersrand hostels. The construction of the *"ideal compounds"* on the Rand was based on that of the diamond mines. Thus, the Kimberley compounds provided a model for compounds on the Rand (Turrell, 1984).

The Kimberley mining hostels initially operated as open compounds from 1871 to 1884. In these open compounds, workers were free to spend their leisure time as they pleased. However, with the shift to underground mining beginning in 1885, the system changed to closed compounds. These closed compounds isolated the workers from the surrounding town and imposed strict controls on their movement to prevent illicit activities.

Before the introduction of closed compounds, De Beers Consolidated Mines Limited (De Beers CML) began discussions with the Cape government in 1884 about the terms under which they would build their convict station for the private use of convict labor in their workforce. The closed compounds were designed after the De Beers' convict station, which was built in Kimberley in 1884. These compounds were established in 1885, following the diamond discovery in Kimberley 15 years earlier. "*The perfection of the compound system may be said to have been attained in these convict barracks, or branch goal*," wrote the Mine Inspector in 1885.

Compound managers attempted to replicate the advantages of convict labor with free labor. From the mining companies' point of view, there were five benefits: increased efficiency, discipline, lower labor costs, upholding economies of scale, and prevention of diamond theft. Over time, the only significant difference between free labor in compounds and prison labor in the convict station was the distinction between compulsory and voluntary service.

Closed compounds were more overcrowded than open compounds. Although overcrowding was also an issue in open compounds, each compound had its own unique characteristics. Some mining companies maintained what they considered "*acceptable death rates,*" while others faced public outrage. Apart from routine visits from the Sanitary Inspector, there was little official regulation of compounds. Once De Beers CML monopolized production, municipal authority was severely limited. Furthermore, monopoly production resulted in the centralization of compound accommodation (Turrell, 1984).

In addition to being overcrowded, compounds were also cold. Workers slept on the ground in quarters without doors and with an open space around the top of the walls. While this aided ventilation during the summer and the day, it did not exclude the intense cold at night and during the winter. Kimberley is notorious for its extreme temperature fluctuations. It has been recorded that it was so cold during the nights that mineworkers were in the habit of going down into the mines early in the afternoon because it was warmer than in their compound (Turrell, 1984).

Closed compounds functioned as self-contained institutions. Sustenance was sold in compound stores, and alcohol was eventually prohibited. Workers traveled between mines and compounds via enclosed subways. The large compounds were covered with a fine wire mesh to prevent stolen diamond parcels from being thrown over the fences. This completed the total isolation of the workers in their world (Turrell, 1984).

To explain closed compounds simply in terms of preventing theft is to abstract one factor from a complex process of class struggle (Turrell, 1984). It conflates a functional explanation with a causal explanation. It disregards the historical period in which closed compounds were first introduced. It should be noted that closed compounds were introduced during underground mining development, which required significant capital investment (Turrell, 1984). The rapid concentration and centralization of capital, demanded stricter control over migrant labor, as its unregulated fluctuations in availability had long hindered the process of capital accumulation.

Most importantly, the closed compound system was regarded as the most effective way for mine owners to deal with migrant labor. The mine owners valued closed compounds because they provided a steady supply of mine laborers. For an industrializing South Africa, closed compounds were an early model of labor control (Turrell, 1984).

The mining hostels were perceived to be a roaring success from the mining company's perspective. Therefore, the same model was used to construct the compounds for the Witwatersrand gold mines. The Witwatersrand open compounds quickly took on the appearance of the Kimberley prototype but lacked the watchtowers (Demissie, 1998). The mining compounds and migrant labor hostels, vital system components during the late colonial and apartheid periods, were designed (and functioned) as tools of control and repression. Over time, they became synonymous with violence, overcrowding, and uncleanliness (Vosloo, 2020). According to Cooke (2007), an examination of the system shows: "*…clearly how colonial and apartheid regimes employed the spatial devices of jails or concentration camps to keep labor present and submissive, and in connection with social regulation, created a divided and violent land.*"

To explain how solid the undercurrent of power runs throughout the mining industry concerning the mining compounds, Vosloo (2020) draws on the work of Michel Foucault. Foucault identified four design tactics through which authorities can gain power in institutions such as prisons, asylums, and schools. The relationship between visibility, control, and power is crucial to his theory.

The first of these architectural strategies is the necessity to construct observatories through the building's layout and spatial configuration. Spaces are built, and buildings are arranged so that occupants can be seen to learn more about them. As a result of their awareness of being observed, they change their behavior through self-policing. By structuring areas this way, the physical environment creates a relay system through which the authorities can exert authority over the inmates

(Foucault, 1977). The power is internalized through self-policing, and the individuals exercise control over themselves without external supervision (Foucault, 1977). Foucault (1977) explained that panoptical layouts are the most obvious way to achieve this goal.

This goal can be attained through strict routines combined with punishment and incarceration. According to Foucault's theory, it is necessary to impose a disciplinary regime and a rigorous program that governs the order in which things are done; even the behavior and sequence of body motions must be regimented (Foucault, 1977). Finally, disciplinary authority is established through examination, which entails a combination of hierarchical observation and standardization of judgment. Vital to note is that these power-embedded strategies were used to regulate the workforce in the migrant labor hostels in South Africa (Vosloo, 2020).

In 1955, Hendrik Verwoerd, Minister of Native Affairs under the National Party played a pivotal role in advancing apartheid policies, including the establishment of the infamous single-sex hostels for mine workers (Magubane & Smith, 2001). This meant that men were removed from their homes for long periods, and the visits by their spouses were also controlled (Bezuidenhout & Buhlungu, 2011). These male-only compounds allowed for offensive reproductive control, which involved conditional rules under which husbands and wives could interact (Bezuidenhout & Buhlungu, 2011).

The migrant labor system had, to a large extent, become synonymous with what people regarded as miserable living conditions in mining hostels (Pelser, 2003). Like its Kimberley counterpart, living conditions in the Witwatersrand single-sex hostels involved poor sanitation, only cold water, concrete bunks for beds, and no privacy due to overcrowding (Magubane & Smith, 2001). Bezuidenhout and Buhlungu (2011) explained that mining companies controlled the miner's nutritional intake and supply, managed the compounds' entertainment spaces, and provided the infrastructure for mine stores. Nutrition was often of low quality and led to protests (Bezuidenhout & Buhlungu, 2011).

Kwamashonisa, or concessioner mining stores, held a monopoly over goods like clothing, radios, and bicycles, dominating the market for consumer items (Bezuidenhout & Buhlungu, 2011). Mining companies controlled all leisure activities within these compounds. Because workers' movements were restricted and regulated, mining corporations effectively had a captive market, enabling them to control consumption. Workers did not have a free choice of destinations or modes of transportation in most situations, nor did they have a free option of where they worked. Also, it was illegal to quit a contract (Massey, 1983). The Employment Bureau of Africa (TEBA) offices aided in investigating workers who had violated their contracts (Bezuidenhout & Buhlungu, 2011).

These dire conditions continued until the early 1980s. During most of the 1980s, resistance toward the treatment mining companies reserved for their Black workers proliferated. It was through the formation of the National Union of Mineworkers (NUM) in 1982 that resistance gained momentum. During that period, mining companies were also starting to rethink the migrant labor system (Bezuidenhout & Buhlungu, 2011). This was mainly due to international pressure. However, it was only during post-apartheid that legislation requiring a degree of corporate social responsibility was developed and institutionalized.

Earlier Mine Workers' Health and Safety

The mining industry in South Africa has a nasty history. Under a wage-constraint system that capped wages to prevent competition between mines for Black workers, brute force and systematic assaults have ensured continued underground production (Stewart et al., 2019). Besides the exploitive labor system, the mining industry is known for its poor role in keeping its workers healthy and safe.

The working conditions of that time were notoriously dangerous. In 1901, the Transvaal Government Mining Engineer observed that mining was more dangerous than war, but the miners got no medals (Van Onselen, 2019). In 1916, the first year for which we have detailed data for South Africa, 8162 miners were killed in accidents, with an accident death rate of 1.56 per 1000 workers. Also, in 1960, it was reported that since 1900, more than 36,000 workers had been killed in gold mine accidents (Simons, 1960). In the same year (1960) it was further reported that the annual death toll from mine accidents in South Africa was around 800. According to Simons (1960), mining accidents in Great Britain killed fewer than 200 people annually. Moreover, an uncountable number of workers died due to septicemia and other diseases caused by accidents. Many others have lost limbs, eyes or have been rendered disabled in other ways.

Besides mining accidents, "*three other serial killers*," as van Onselen (2019) calls it, were prevalent: pneumonia, silicosis, and tuberculosis (TB). Pneumonia, silicosis, and TB did not exist in isolation. There was a cross-cutting of influences at work for all Black miners, especially deep-level miners (Van Onselen, 2019). Alcoholism, inadequate mine diets, exhaustion, and physical stress added to the high accident rate and the migrant miners' reduced resistance to other diseases (Van Onselen, 2019). To illustrate how inhumane the conditions were, an estimation of the number of deaths of all Black mineworkers on the Rand between 1902 and 1912 was estimated at 50,000. To put this in perspective, van Onselen states that this loss of life is equivalent to the sinking of 33 Titanics.

Still, these figures should be treated as estimates, and a great deal of caution should be practiced when considering the number of deaths of mineworkers during the early 1900s. A reason for making the figures unreliable was the migrant labor system. The number of men who became ill or died in the mines was recorded in official data. There was, however, always a disparity between this figure and the high mortality rates. Men who died in accidents and those who died in the compounds from infectious diseases such as pneumonia, meningitis, or typhoid fever were recorded, but silicosis and tuberculosis can take years to kill, and most deaths occurred in rural areas, out of sight. Furthermore, the mining industry repatriated sick and dying men to reduce compensation costs (McCulloch, 2016).

Also, some diseases spread to rural communities and across national borders because most miners were migrant workers. Since the 1920s, there has been strong evidence that South Africa's mines have spread tuberculosis to neighboring states (McCulloch, 2016). Furthermore, if the gold industry had drawn its workforce from the surrounding towns and villages, the towns and villages would have quickly filled with men crippled by occupational lung disease. As a result, political pressures to reduce risks would have increased. The High Commission Territories of Bechuanaland (Botswana), Swaziland, and Lesotho, significant sources of labor, had no X-ray facilities and specialists to identify the disease burden among returning miners.

The Mining Industry and Reconciliation

Examining the South African mining industry within a broader global context underscores a fundamental truth about corporate behavior, revealing that it conforms to a prevailing pattern. This perspective acknowledges that the industry is not an isolated exception but rather a reflection of the broader tendencies exhibited by corporations.

Joel Bakan's renowned book, *The Corporation: The Pathological Pursuit of Profit and Power* (2004), and its corresponding documentary brought into focus the role of businesses in society. Bakan argued that the very structure of corporations, designed to prioritize profit above all else, is a chief contributor to the ethical and legal issues that permeate the commercial world (Bakan, 2004). By its nature, the corporation shields its leaders and owners from legal accountability, resulting in an entity with a "*psychopathic contempt for legal constraints*" (Bakan, 2004). This understanding serves as a lens through which we can explore the actions and behaviors of the South African mining industry in a broader global context.

Contrasting views around a corporation's role in the twenty-first century have also emerged. In opposing opinions, corporations, specifically multinational corporations, are regarded as the "*...backbone of our current global web of institutions that deserve praise*" (Horrigan, 2010). Webster et al. (2008) recognize that while the global economy has created opportunities for wealth creation, it simultaneously threatens established livelihoods (Webster et al., 2008). Webster et al. (2008) indicated that the ecological harm caused by the market emerged as the primary central social issue of the twenty-first century.

McKibben (1989) regarded multinational corporations as environmental destroyers. Besides having severe impacts on the environment, multinational corporations may also violate human rights. In this context, corporations' social responsibility becomes essential to businesses due to the widespread destruction of human rights and the environment (Haque & Azmat, 2015). The persistent human rights violations by corporations in recent decades prompted the United Nations Human Rights Council to universally adopt the Guiding Principles for Business and Human Rights in 2011, highlighting the urgent need for corporate accountability and ethical business practices.

The history of South Africa's mining industry bears a dark legacy of human rights violations that continues to cast a long shadow. By the end of apartheid, the impacts of mining hostels and the migrant labor system, along with the various social issues they created, were deeply embedded in South Africa's socioeconomic landscape. So serious were these socioeconomic fault lines that the post-apartheid Truth and Reconciliation Commission (TRC)[1] spoke out a judgment over the mining sector. The TRC labeled the mining industry's role as "*first-order engagement*" with the apartheid state (Rapoport Center for Human Rights and Justice, 2019). For the TRC (1998), the Chamber of Mines' actions in shaping the migrant labor system is*:*

> ... the clearest example of business working closely with the (minority) White government to create the conditions for capital accumulation, based on cheap African labor ... the mining industry harnessed the services of the state to shape labor supply conditions to their advantage. Thus, the mining industry bears a great deal of moral responsibility for the migrant labor system. (TRC, 1998: Section 63)

Furthermore, the TRC found that the mining industry played a crucial role in shaping and driving cheap labor policies through various tactics. These tactics are as follows:

- Influencing legislation that forced Black workers into the wage system (and managed their allocation within it).
- State-endorsed monopolistic recruiting practices.
- The capping of African wages.
- Divisive labor practices in managing compounds.
- The brutal repression of Black workers and trade unions.

The TRC concluded that while mining is inherently dangerous:

> …there appears to be some evidence that profitability ranked higher than people's lives—as evidenced by the asbestos scandal and the continued use of polyurethane in mines long after the dangers became known.

Critics argue that the TRC downplayed the economic exploitation of apartheid, treating it as a secondary issue and separating it from the major human rights abuses it focused on. For some, the TRC failed to recognize the mining industry's role in supporting apartheid as a specific human rights violation. To explain this further, the Benchmarks Foundation (BMF), submitted in 1998 the following explanation:

> The human rights violations by business are seen as those policies, practices, and conventions which denied Black people the full utilization of their potential, resulting in deprivation, poverty and poor quality of life, and which attacked and threatened to injure their self-respect, dignity and well-being. Certain of these violations were open abuses, whilst some were indirect; yet others buttressed those carried out at a socio-political level.

[1] The Truth and Reconciliation Commission (TRC) was a judicial-like entity formed in 1995 to examine and tackle human rights violations that took place during apartheid in South Africa.

Also, in 1998, the African National Congress (ANC) submitted that:

> It is our contention that the historically privileged business community as a whole must accept and acknowledge that its current position in the economy, its wealth, power and access to high-income and status positions are the product, in part at least, of discrimination and oppression directed against the Black majority.

Given this entrenched racialized inequality, economic justice in the mining sector was a vital objective of the liberation struggle and the democratic transition. For this reason, after 1994, with the fall of apartheid, legislation was put in place to address the myriad of injustices from the past and, among other things, to specifically foster artisanal and small-scale mining (ASM) in South Africa.

For the mining industry, the key piece of legislation that was developed to create opportunities for historically disadvantaged South Africans is the Mineral and Petroleum Resources Development Act (MPRDA) of 2002 (Moraka & Van Rensburg, 2015).

Designed to support ASM, the Act paradoxically prioritizes large-scale and mechanized mining instead. The permitting process for small-scale mining operations under 5 ha, within a 2-year timeframe, is outlined in Section 23 of the Act (SAHRC, 2015). Small-scale miners are exempt from submitting social and labor plans (SLPs), unlike their larger counterparts, but they are still required to provide an Environmental Management Plan to the Department of Mineral Resources and Energy (DMRE). However, the substantial skills, capacity, and financial resources required to navigate the permitting process create a significant barrier for those seeking to mine legally on a small-scale basis (Debrah et al., 2014).

This mining permit is the sole legal pathway for aspiring small-scale miners, yet there are no alternative permits, rights, or licenses available for recognition as a small-scale miner (Legal Resources Centre, 2016). For ordinary South Africans and for those in the Zama Zama mining sector, the technical and financial requirements to obtain a mining permit are out of reach. In the end, the Act makes insufficient room for artisanal mining by focusing on small-scale mining rather than on artisanal mining (Debrah et al., 2014; SAHRC, 2015). Consequently, marginalizing and criminalizing the artisanal mining sector effectively.

Conclusion

Mining has been an essential part of the South African economy for almost 150 years, resulting in unequal socioeconomic harm and gains (Rapoport Center for Human Rights and Justice, 2019). Also, since the start of large-scale gold mining in South Africa in the late 1800s, little effort has been made to include artisanal miners in the country's economic development. This exclusion created an equity gap.

Only in the post-apartheid period, after 1994, was legislation first put in place to foster ASM in South Africa. However, legislation primarily focuses on small and industrialized mining, making no room for any artisanal type of mining, consequently excluding historically disadvantaged people from the mineral riches of South Africa.

The sector is either ignored or suppressed by powerful actors like mining companies and the government. By failing to adequately legally accommodate artisanal miners, they are effectively marginalized and classified as illegal. It is essential to acknowledge that artisanal mining in South Africa has developed, and continues to occur, within this socio-legal context.

Having outlined the historical context of the broader mining industry and how artisanal mining came to position itself in this context, the next chapter will delve into an in-depth examination of artisanal mining in South Africa, highlighting its distinct dynamics and challenges.

References

Bakan, J. (2004). *The corporation*. Constable and Robinson.

Bezuidenhout, A., & Buhlungu, S. (2011). From compounded to fragmented labour: Mineworkers and the demise of compounds in South Africa. *Antipode, 43*(2), 237–263. https://doi.org/10.1111/j.1467-8330.2010.00758.x

CJPME Foundation. (2014). *Roots of apartheid: South Africa's mining industry*. Accessed December 20, 2021, from Roots of Apartheid: South Africa's Mining Industry - The CJPME Foundation

Cooke, J. (2007). The form of the migrant labour hostel. *Architecture South Africa, July/August*, 64–69.

Crush, J., Jeeves, A., & Yudelman, D. (1991). *South Africa's labor empire. A history of black migrancy to the gold mines*. David Philip Publishers.

Demissie, F. (1998). In the shadow of the gold mines: Migrancy and mine housing in South Africa. *Housing Studies, 13*(4), 445–469. https://doi.org/10.1080/02673039883218

Debrah, A. A., Watson, I., & Quansah, D. P. O. (2014). Comparison between artisanal and small-scale mining in Ghana and South Africa: Lessons learnt and ways forward. *The Journal of the Southern African Institute of Mining and Metallurgy, 114*, 913–921. http://www.scielo.org.za/pdf/jsaimm/v114n11/09.pdf

Foucault, M. (1977). *Discipline and punish: The birth of the prison*. Penguin.

Haque, M. Z., & Azmat, F. (2015). Corporate social responsibility, economic globalisation and developing countries. *Sustainability Accounting, Management and Policy Journal, 6*(2), 166–189.

Horrigan, B. (2010). *Corporate social responsibility in the 21st century. Debates, models, and practices across government, law and business*. Edward Elgar.

Legal Resources Centre. (2016). Artisanal Mining Report. http://lrc.org.za/wp-content/uploads/pdf/2016_Artisanal%20Mining%20Report_LRC%20Publication.pdf. Accessed 25 April 25 2018.

Magubane, P., & Smith, C. (2001). *Soweto*. Struik.

Massey, D. B. (1983). Class struggle and migrant labor in South African gold mines. *Canadian Journal of African Studies, 17*(3), 429. https://doi.org/10.2307/484926

McCulloch, J. (2016). Sleights of hand. *New Solutions: A Journal of Environmental and Occupational Health Policy, 25*(4), 469–479. https://doi.org/10.1177/1048291115610432

McKibben, B. (1989). *The end of nature*. Random House.

Meredith, M. (2007). *Diamonds, gold, and war: The making of South Africa*. Jonathan Ball Publishers.

Moraka, N. V., & Van Rensburg, M. J. (2015). Transformation in the South African mining industry - looking beyond the employment equity scorecard. *Journal of the South African Institute of Mining and Metallurgy, 115*(8), 669–678. https://doi.org/10.17159/2411-9717/2015/v115n8a2

Pelser, J. (2003). Migration in South Africa: A profile of patterns, trends and impacts. In L. Adler & U. Gielen (Eds.), *Migration: Immigration and emigration an international perspective*. Praeger Publishers.

Rapoport Center for Human Rights and Justice. (2019). *Property rights from above and below: Mining and distributive struggles in South Africa*. Accessed December 24, 2022, from https://law.utexas.edu/wp-content/uploads/sites/31/2019/12/Report_Mining-and-Distributive-Struggles-in-South-Africa-December-2019.pdf

Simons, H. J. (1960). *Death in South African mines*. Accessed December 19, 2021, from asjul61.8.pdf (sahistory.org.za)

South African Human Rights Commission (SAHRC). (2015). Report of the SAHRC Inves-tigative Hearing Issues and Challenges in Relation to Unregulated Artisanal Underground and Surface Mining Activities in South Africa. https://www.sahrc.org.za/home/21/files/Unregulated%20Artisanal%20Underground%20and%20Surface%20Mining%20Activities%20electronic%20version.pdf. Accessed 12 November 2015.

Stevenson, B. (2014). *Just mercy*. Random House Publishing Group.

Stewart, P., Bezuidenhout, A., & Bischoff, C. (2019). Safety and health before and after Marikana: Subcontracting, illegal mining and trade union rivalry in the South African mining industry. *Review of African Political Economy, 47*(163), 27–44. https://doi.org/10.1080/03056244.2019.1679103

Truth and Reconciliation Commission of South Africa. (1998). *Truth and Reconciliation Commission of South Africa Report* (Vol. 4, Chapter 2). Retrieved from Justice.gov.za

Turrell, R. (1984). Kimberley's model compounds. *The Journal of African History, 25*(1), 59–75. https://doi.org/10.1017/s0021853700022568

Van Dyk, M. C. (2018). *Race, resistance and the general tax of 1925: A historical overview of the interpretation and implementation of South African poll tax*. University of South Africa. http://hdl.handle.net/10500/25336

Van Onselen, C. (2019). *The night trains*. Jonathan Ball Publishers.

Vosloo, C. (2020). Extreme apartheid: The South African system of migrant labour and its hostels. *Image & Text, 34*. https://doi.org/10.17159/2617-3255/2020/n34a1

Webster, E., Lambert, R., & Bezuidenhout, A. (2008). *Grounding globalisation: Labour in the age of insecurity*. Blackwell.

Wilson, F. (1972). *Migrant labour: Report to the South African Council of Churches*. Ravan Pr of South Africa.

Chapter 4
Artisanal Mining in the South African Landscape

> *Poverty is not only about income poverty, it is about the deprivation of economic and social rights, insecurity, discrimination, exclusion and powerlessness. That is why human rights must not be ignored but given even greater prominence in times of economic crisis.*
>
> —Irene Khan (as cited by The Borgen Project n.d.)

Introduction

The artisanal and small-scale mining (ASM) taxonomy has many definitions. This chapter explores the complex world of artisanal mining in South Africa by untangling the intricacies of small-scale and artisanal mining. These two terms are often used interchangeably but carry distinct implications for the communities engaged in these activities. Understanding this distinction lays the foundation for a meaningful investigation into the artisanal mining sector.

A challenge in the artisanal mining sector is the scarcity of comprehensive data. This chapter unpacks the data gap surrounding artisanal mining. This data gap has hindered the ability to fully grasp artisanal mining's influence on economies, communities, and the environment. Furthermore, the chapter explores the question of whether artisanal mining helps people escape poverty or traps them in it. Is it a desperate last resort for the marginalized, or does it offer a means of economic empowerment? These questions lie at the heart of the artisanal mining debate.

The chapter subsequently looks at the nature and activities of the Zama Zamas, shedding light on their profile and the specific minerals they extract in South Africa. Lastly, we will examine the relationship between large-scale mining (LSM) and the rise of artisanal mining, exploring how the actions and dynamics of the former drive the persistence of Zama Zama mining operations.

V. Bester, *The Untold Story of Zama Zama Miners in South Africa*, Clinical Sociology: Research and Practice, https://doi.org/10.1007/978-3-031-82783-9_4

The Distinction Between Small-Scale and Artisanal Mining

Artisanal mining is often lumped together with small-scale mining under the collective term ASM. There is no general agreement on what constitutes small-scale and artisanal mining, and the borders between artisanal and small-scale operations are often unclear (United Nations Economic Commission for Africa, 2011). This is because definitions differ from country to country, with each country outlining its criteria, entailing different parameters around small-scale and artisanal mining.

Not clearly distinguishing between the two sectors is problematic since it creates a misunderstanding about artisanal mining. Consequently, solutions to address the sector often miss the mark. To provide tailored solutions, there is a need to contextualize the differences between ASM (Legal Resources Centre, 2016). As an alternative, artisanal and small-scale miners are often viewed on a scale of mining activities, fluctuating from individuals panning for gold to relatively more extensive, mechanized, and organized operations (Collins & Lawson, 2014). When viewing ASM as part of a range of mining activities, the difference between ASM stays blurred without distinguishing between the two sectors.

The challenge is no different in South Africa, where artisanal mining is often clustered together with small-scale mining under the collective term ASM (Legal Resources Centre, 2016). At first, the former Department of Mineral Resources[1] (DMR) referred to artisanal mining under the umbrella term "small-scale mining" without any differentiation. The then-DMR (2017) defined the following subgroups under small-scale mining: *"Artisanal or subsistence mining operations (new entrants); sub-optimal formal mining operations; and entrepreneurs with upfront capital."*

However, the then-DMR did not describe each sub-category in detail. Moreover, what differentiates the sub-categories from each other is not clear.

The National Small Business Amendment Act, No. 26 of 2003, has occasionally been used in an attempt to define artisanal mining. The Act applies the categories of annual turnover and employment numbers to outline small business activities in the mining and quarrying sector. The classification of operations is depicted in Table 4.1.

In principle, these measures create a range of small-scale mining activities ranging from medium to micro, which is all good and reasonable. However, in practice, the above categorization falls short. A mining operation or business may conform to one condition and, at the same time, may not automatically fit into the other specifications. In short, this classification system does not reflect the complexity of the differentiation required in ASM in South Africa (Mutemeri & Petersen, 2002).

South Africa's ASM policy, first released in May 2021 as a draft policy, proposes two distinct definitions to address this issue. The policy identifies the main

[1] In 2019, the Department of Mineral Resources and Energy (DMRE) was created in June 2019 when the Department of Mineral Resources and the Department of Energy were merged; previously, the DMRE was known as the Department of Mineral Resources (DMR).

Table 4.1 Classifications of mining and quarrying operations per the National Small Business Amendment Act, No. 26 of 2003 (South African Government, 2003)

Size	The total full-time equivalent of paid employees	Total turnover
Medium	200	ZAR 39M
Small	50	ZAR 10M
Very small	20	ZAR 4M
Micro	5	ZAR 200,000

characteristics of artisanal mining as related to using elementary mining methods. The policy describes artisanal mining as follows (DMRE, 2022):

> Artisanal mining means traditional and customary mining operations using traditional or customary ways and means. This includes the activities of individuals mainly using rudimentary mining methods, manual and rudimentary tools to access mineral ore, usually available on the surface or at shallow depths.

In contrast, small-scale mining is defined in terms of what it is not:

> Small-scale mining means a prospecting or mining operation that does not employ specialized prospecting, mechanized mining technologies, chemicals including mercury and cyanide or explosives, or the proposed prospecting or mining operations that do not involve investment or expenditure which exceed such amount as may be prescribed.

Complimenting the definition by the Department of Mineral Resources and Energy (DMRE) of artisanal mining, the Legal Resources Centre (2016) describes artisanal mining as a type of mining on a small scale. It occurs mostly informally with hand tools and a high degree of manual labor (Legal Resources Centre, 2016). Artisanal mining does not use machinery, which is often used in organized small-scale mining (Legal Resources Centre, 2016).

Derived from these definitions, using rudimentary tools as part of a basic process becomes a distinguishing characteristic of artisanal mining. The following paragraphs, accompanied by photographs, detail how Zama Zamas extracts gold.

The Zama Zamas enter reopened holes from abandoned or derelict mine sites to find gold-bearing ore. The artisanal miners spend days and sometimes weeks underground, to find ore-bearing rocks. Pieces of rock are then brought up from the underground tunnels. Then the rocks get crushed with a tool called a *penduka*. A *penduka* is a cylinder, often an old gas bottle, with iron balls inside that is turned for hours to crush the ore (see Fig. 4.1).

The crushed material then gets washed, as seen in Fig. 4.2. After crushing and washing the ore, mercury is mixed with a type of slush containing gold. The mercury combines with the gold, and an amalgamate is formed, consisting of a half-half mixture of gold and mercury.

The amalgamation process can be seen below in Fig. 4.3. The mercury is then burnt away with a blow torch, leaving only the gold behind.

Fig. 4.1 Zama Zamas busy turning a *penduka* (Bester, 2019)

Figure 4.4 shows a typical artisanal mining site where all the above-described phases occur.

Artisanal Mining and the Data Gap

As of 2024, over 225 million people are directly and indirectly involved in the sector's labor value chain across Latin America, South America, Africa, and Asia (World Bank, 2024). According to the World Bank, the total workforce in the ASM sector was estimated at 44 million in 2020 (World Bank, 2020). This number of workers far exceeds the total number of employees in the industrial mining industry, which amounts to 7 million employees (ILO, 2019).

Furthermore, the ASM numbers show a continuous increase. The number of ASM operators in Africa in 1999 stood at nearly 2 million in 24 African countries. Fifteen years later, in 2014, ASM workers stood at almost 10 million (The International Institute for Sustainable Development, 2017). Today, ASM is taking place in more than 80 countries and serves as the backbone of local economies in several countries.

ASM activities typically occur in rural and impoverished areas and low and middle-income countries (Barreto et al., 2018). However, any numbers applied to this sector are probably underestimated, and the sector is likely to play an even more significant role in rural and poor areas where limited economic activity occurs (de Haan et al., 2020). Consequently, ASM is integral in impoverished rural communities in South Africa (Action Aid, 2019).

Fig. 4.2 Zama Zamas washing the ore for the drainage process to begin (Bester, 2019)

To further illustrate the sector's scale, it is worth considering that ASM is responsible for 25% of the global diamond supply, 20% of the global gold supply, and 80% of global sapphires. The sector supplies 18–30% of the world's cobalt.

Interestingly, cobalt is crucial in battery metal powering for (clean) energy (World Bank, 2020). One can only predict that the demand for cobalt will become more significant, with fossil fuel taking a back seat, as the rush for green energy minerals increases. Scholars have rightly remarked that despite the contributions to global mineral production, the ASM workers are predominantly marginalized, and their economic contribution is primarily neglected (World Bank, 2020; Hilson & McQuilken, 2014). However, as alluded to, many of the world's strategic metal supply

Fig. 4.3 Mercury, gold, and water: the amalgamation process (Bester, 2019)

chains for clean energy, information technology, and jewelry rely on materials produced by artisanal miners (Perks & Schneck, 2021).

The ASM sector falls mainly within the informal economy. Despite the sector's informal and often illegal nature and primarily unmechanized operations, it represents a meaningful livelihood for people directly involved and those indirectly dependent on the proceeds of this sector. Considering the figures mentioned, it is essential to acknowledge that an overall worldwide data gap around ASM exists. The World Bank's State of the ASM sector report (World Bank, 2020) explains that this data gap is defined by the near total absence of national and sub-national baseline exercises that collect qualitative and/or quantitative data, and the corresponding lack of data shared collectively in the public domain.

Fig. 4.4 A typical Zama Zama site (Bester, 2023. Unpublished)

Regarding a lack of data, South Africa is no different in this matter. Several documents have reported lacking detailed qualitative and quantitative data on South Africa's ASM sector (SAHRC, 2015; UBC and Canadian International Resources and Development Institute (CIRDI), 2019; Action Aid, 2019). A consequence of a lack of data is that it fuels an absence of stakeholders' understanding of the current realities in the ASM sector. It has been stated that journal articles and research reports provide insight into the state of the ASM sector. However, information gathered through direct interaction with artisanal miners and the communities from which they originate or are accommodated is as effective, if not more so (CIRDI, 2019).

The absence of data means that the possibility for meaningful dialogue, effective strategy formulation, and the identification of necessary policy and legislation reforms remains far off (CIRDI, 2019). Thus hindering development from a multi-stakeholder perspective (CIRDI, 2019).

What adds to the situation's complexity is that many miners in the ASM sectors operate outside the legislative realm. Therefore, available data regarding the ASM sector in South Africa, as housed by the DMRE, is only on the legal, small-scale mining side of the spectrum and ignores the complex and economically active artisanal miners on the far end of the spectrum. Despite significant data gaps, some estimates regarding the number of Zama Zama miners have been made.

The total number of artisanal or Zama Zama miners in South Africa is estimated to be 30,000 (SAHRC, 2015). However, The Bench-Marks Foundation (2017) projected 30,000 artisanal or Zama-Zama miners in the Gauteng Province alone. The Legal Resources Centre (2016) reports 14,000 artisanal gold miners operating in approximately 6000 abandoned mines. Moreover, the number of dependents is even more significant; researchers have estimated that 250,000 people survive on the proceeds of Zama Zama mining (Bench-Marks Foundation, 2017). It is crucial to treat these numbers with caution for two reasons. Firstly, the 2015 South African Human Rights Commission (SAHRC) estimation was a decade ago, making the figure somewhat outdated. Secondly, the numbers are likely higher considering the ongoing mine retrenchments, mine closures, and the socioeconomic impacts of the COVID-19 pandemic. Consequently, the number of dependents is also likely to be much higher.

The lack of visibility of the Zama Zamas in the data further underscores how overlooked this sector has been for years. Fueling this is that the operations are illegal and do not take place in the realm of law. On this note, the CIRDI (2019) says that reliable qualitative and quantitative data on the illegal gold mining sector is difficult to find.

Accurate empirical-based information will allow for the creation of nuanced and multidimensional policy responses and practices. In the end, this could contribute to halting the criminalization of the sector and enable it to develop socioeconomically. One way to do this is to involve multiple stakeholders like the Zama Zamas, civil society, and government representatives in a dialogue, that can contribute to accurate and fresh information about Zama Zama mining.

Artisanal Mining: Poverty Trap or a Way of Making a Living?

Artisanal mining is widespread in South Africa and is an income-generating stream for the unemployed. South Africa has one of the highest unemployment rates globally, with an official rate of 33.5% in the second quarter of 2024 (StatsSA, 2024). Additionally, in the same quarter of 2024, Gauteng's provincial unemployment rate was 35.1% (StatsSA, 2024). Therefore, finding alternative means of livelihood is crucial.

However, many scholars have described the ASM sector as a poverty trap. Some narratives, as Hilson (2009) explains, describe ASM in sub-Saharan Africa as a "*rush-type*" activity. This can make it chaotic and entrepreneurial-driven, where miners are regarded as "*fortune-seekers*". Sometimes, ASM is also seen as a "*distress-push*" activity where miners seek to alleviate their poverty and work in ASM to complement revenues from farming (Hilson, 2009). For example, Hilson (2016) discusses the seasonal nature of ASM and farming–ASM linkages in sub-Saharan African countries such as Ghana, Liberia, Malawi, Mali, Mozambique, Sierra Leone, and Zimbabwe. In these cases, ASM supplements farming activities.

When developments at ASM sites in sub–Saharan Africa are viewed through a temporal lens, it becomes clear how the poverty-driven and platform for wealth creation narratives are inextricably linked. Also, ASM is critical to poverty reduction in rural areas of sub-Saharan Africa and contributes significantly to household livelihoods. As mentioned, in some cases, the ASM sector is a supplementary income source for rural households (Hilson, 2016).

Acknowledging the sector's potential to absorb the unemployed youth is also important. Careful consideration is essential, as the youth are frequently mistakenly associated with child labor within the sector (De Haan et al., 2020). Consequently, to uphold the argument for youth employment in the sector, it is imperative to establish a distinct differentiation between young individuals engaged in the sector and child labor.

South Africa had a youth unemployment rate (aged between 25 and 34 years) of 41.7% during the second quarter of 2024 (StatsSA, 2024). This is motivation enough to support sectors that can absorb the unemployed youth. Action Aid reported that Zama Zamas, who fall under youth in the South African context (e.g., between 18 and 34 years), account for 38% of the respondents engaged in artisanal mining (Action Aid, 2019). The youth's participation in this sector indicates that this neglected socioeconomic sector can help to address youth unemployment in South Africa, provided creative and practical ways are designed and implemented to manage the sector.

However, much skepticism exists around ASM's potential to reduce poverty (De Haan et al., 2020). Supporters of this argument reason that the ASM sector consists of unskilled people. The unskilled artisanal miners' workforce relies on rudimentary practices due to a lack of support and meager capital. Eventually, these elementary practices cause low productivity and low efficiency with little income (De Haan et al., 2020). Also, ineffective mining practices cause environmental damage, ill health, and social issues, trapping miners in a cycle of poverty. A counterargument to this is that many artisanal and small-scale miners prefer remaining in the ASM sector (De Haan et al., 2020). According to this reasoning, ASM is the most promising livelihood strategy for escaping rural poverty.

Both these arguments are valid, and both carry circumstantial truth. To illustrate this and the complexity of the matter, the 2019 study on Zama Zama mining conducted by Action Aid, in South Africa states the following: Most of the respondents in Vlakfontein and Carletonville, both mining towns in South Africa, said that they would leave the artisanal mining sector if they obtained alternative employment. That is 92% and 70%, respectively. In Kimberley, another South African town, most respondents (60%) indicated they would not leave artisanal mining even if they obtained other jobs. Those who answered *"somewhat"* (3% in Kimberley and Carletonville and 6% in Vlakfontein) indicated that it would depend on the earnings offered. Also, during my discussions with Zama Zamas, many said they would leave the sector should the alternative livelihood be more feasible.

Given this context, one might wonder: What factors cause the artisanal miners to provide such conflicting responses? It might be about the gold and diamond prices since diamonds are mined in the Kimberley area. Alternatively, it might be about the broader social structure of each setting, the presence of syndicates, or the turf wars that do break out among certain Zama Zama factions. Other than theorizing about the reason for such opposing answers, the main (disguised) component here is the heterogeneous nature of the sector, which should, if anything, make one wary of treating the whole sector the same.

Exploring the nuances of the sector will allow for a deeper understanding of the sector. Comprehensive knowledge of the artisanal mining sector in South Africa is much needed since, if adequately supported, the sector can serve as a powerhouse of economic development in poor communities (Mutemeri & Petersen, 2002).

On the surface, especially if the media accounts are being followed, it will look like artisanal miners are motivated by mere criminal intent and will go out of their way to cause havoc and violence. However, by taking a closer look at the matter, it becomes clear that this is not the case and that entertaining a homogenous view is, in fact, erroneous.

Who Are the Zama Zamas?

During many conversations, it was mentioned that the generic artisanal mining profile is firmly based on misconceptions. The interviewees said this creates a one-dimensional view of Zama Zama mining in South Africa. A civil society worker said the following:

> "I must also just say that there are different kinds of micro and survival mining operations. And people usually say that Zama Zamas is a blanket term, but they don't differentiate between the different categories of groups of people that there are."—Danny, civil society

Interviewees explained that there are many variables among the Zama Zama miners. Firstly, the type of minerals artisanal miners mine was mentioned as an element differentiating Zama Zamas group*s*.

> "Well, I think to some extent that coal artisanal mining is very different from gold. Because the coal that people mine just gets sold to people in the townships and so on as energy. Meanwhile, gold gets sold on the international market. I think that illegal mining in diamonds is similar to gold, except that they're, well in Kimberley there are abandoned dumps that people are mining. De Beers doesn't want them to mine, but De Beers is not doing anything about the dumps themselves. In platinum, there are miners, but in platinum, it is not on abandoned mines. It is not on surface mines. It is this syndicated type of operations that I have described earlier."—Danny, civil society

Jane, a member of civil society, had the following to say in terms of the importance of considering the different types of minerals the Zama Zamas mine:

> "From my point of view, the idea that you have a generic profile across all minerals is problematic. Because I think that there are very different profiles and conditions both in terms of the miner, as well as in terms of the geology for different minerals. And there are then also different prospects for whatever is done at the artisanal level, how that's used."

During a discussion with a worker from another civil society organization, it was suggested that a good starting point to unpack the differences between Zama Zamas would be an interrogation of Zama Zama's socioeconomic backgrounds.

> "I think, socioeconomic factors and, and which standpoint you're looking at it from also, has a huge impact on how you would look at a profile of an artisanal miner. And I think people['s] socioeconomic circumstances need to be interrogated."—Nicky, civil society

Considering and investigating the different socioeconomic circumstances of Zama Zamas will acknowledge the heterogenic nature of this sector, where currently, little reference to distinguishing characteristics has been made. The acknowledgment of a heterogenic profile is rarely, if ever, considered. Furthermore, the assumption of a homogenous profile narrows engagement initiatives and fuels policing, further marginalizing the sector. Both consequences, of course, are highly unsustainable.

When interviewees were asked about the differences in the Zama Zama sector, some indicated that there are different types of Zama Zamas. In some conversations, it was suggested that a distinction should be made between Zama Zamas mining in operating mines and those mining in abandoned derelict or ownerless mine sites. Jane, a member of civil society, said:

> "For me, where there have been instances where formal, formally employed miners have used their position of formal employment, to allow access, I think that that's very, very problematic. I mean, I certainly don't... It's not about moral judgement but it's about the extent of the exploitation...it's illegal and exploitative. There's no way to incorporate it into legal processes. I think, I would distinguish from that point of view. Not, not from a moral judgement point of view."

In a similar vein, another member of civil society, Danny, reasoned that differentiating between the different types of Zama Zamas is essential. He commented on the differences between Zama Zamas in abandoned mines and those in operating mines; he said:

> "There are those who occupy abandoned mine sites, and there is usually huge informal settlements on these abandoned mine sites, and they operate these abandoned mine sites. Meaning that, legally speaking, they are not trespassing because the mine site is not owned by anyone. Okay. And the mining rights are not owned by anyone. Then you have what you have in ultra-deep level mines in the Far West Rand and also at places like Welkom. You have formal mining operations in which there is shadow mining taking place and that is where the illegal operators come in, go down with the legal employees of the mine stay underground for six months at a time and send up what they produce through the formal cage system what is going up the shaft. And receive food and payment down the shaft all the time."

What stood out from these conversations was that not all interviewees shared the same viewpoints. Asked if there are any differences between Zama Zamas in abandoned mines and those mining in operating mines, Themba, a member of civil society, said:

"I think that would be location specific. Yeah, it would be location specific for instance in the East Rand most of the gold mines are abandoned gold mines, but if you go more towards Mpumalanga, the Barberton area, there is still some form of formal mining happening, and the Zama Zamas operating side by side."

On the other hand, Mark, a key informant, offered the following:

"I think the only difference is, in abandoned mines they are not often open, they are not bothered. Whereas in existing operations, they're more careful not to be caught, some of those activities would occur at night."

These opposing viewpoints reflect the complexity of Zama Zama mining. Field research also found that some Zama Zamas had an educational background. The following statement by an interviewee indicates how diverse the sector is.

"...some of them I've met geologists and metallurgists who are qualified professional people."—Carina, key informant

The empirical data about the profile of Zama Zamas have important implications for initiatives targeting Zama Zama mining. For one, the views and perceptions of the participants underscored that when initiatives are considered and developed, they should be coupled with thorough baseline research. Socioeconomic differences should be considered when any initiatives or policies are developed. It is imperative not to assume that the entire sector consists of unskilled laborers who will mindlessly follow any alternative livelihood measure. The issue with the misrepresentations of the Zama Zama profile is that they stigmatize and curb developmental initiatives.

One of the issues that emerged from these findings is the existence of different types of artisanal mining. It was indicated that having a generic profile across miners is problematic. This implies the need to uncover socioeconomic factors within the sector.

Moreover, these findings suggest that artisanal mining is complex and that an oversimplification of the profile of miners is erroneous. Scholars have also argued this point when researching informal settlements. Studies on informal settlements have pointed out that *"slums and squatter settlements"* are oversimplified. It has further been argued that effective intervention programs that aim to improve the lives of people living in informal settlements must be based on an accurate understanding at a local and national level (Huchzermeyer & Karam, 2006).

It seems that the informal sector, be it informal settlements or informal mining, is frequently oversimplified and presented rigidly. Informal settlements' changeability and multifaceted relations are mostly lumped into one homogeneous category (Groenewald et al., 2013). The same can be said about artisanal mining. Seldom, if ever, is a distinction made regarding socioeconomic factors, nor is the sector's flexibility recognized.

Where and What Do Zama Zamas Mine in South Africa?

Worldwide, artisanal miners exploit over 35 minerals but concentrate on high-value minerals such as gold, coltan (columbite-tantalum), and precious and semi-precious gemstones like diamonds and sapphires (Toteu, 2017). ASM can include full-time or seasonal miners (Buxton, 2013). When new ore is discovered in an area, an influx of people may occur in a settlement, causing a rapid rise in a community's population over a short period (O'Faircheallaigh & Corbett, 2016).

In South Africa, most artisanal mining occurs in abandoned and disused mines in historic mining communities. However, Zama Zama mining does occur at active mine sites, causing violence and conflict with the security forces of formal mining companies. In 2019, the then and current Mineral Resources Minister Gwede Mantashe reportedly said the illicit mining industry cost South Africa 41 billion ZAR in 2018 alone (ENCA, 2019). It is unclear how these figures are calculated since most Zama Zama mining occurs at abandoned or derelict mine sites as an unregulated activity.

Reports indicate that artisanal mining occurs in all nine South African provinces but is more prevalent in the gold sector, with most of the mining occurring in

Fig. 4.5 Artisanal miners at work (Bester, 2023. Unpublished)

Gauteng, Free State, Mpumalanga, and the North West. Figure 4.5 shows some artisanal miners mining for gold in Gauteng.

The artisanal miners mine other commodities such as coal, chrome, base metals, semi-precious stones, platinum, industrial minerals, and construction materials like sand and clay mining.

In addition, in the Northern Cape, most artisanal mining activities involve diamonds in the Kimberley area, the Namaqualand, and Port Nolloth regions of South Africa. Artisanal miners are also mining tiger's eye deposits in areas surrounding Prieska. It is projected that there are about 20,000 diamond miners in the Northern Cape Province alone. As for the North West Province, the mainstream of artisanal mining activities occurs in the gold sector around Stilfontein and Klerksdorp. In the Mpumalanga Province, artisanal miners exploit coal and gold. Coal mining occurs within the Witbank and Ermelo regions, primarily in disused underground mines, while gold mining is concentrated in the Barberton area. As for the Eastern Cape, sandstone is mined in the mountainous areas of this province (CIRDI, 2019). Coal mining is also dominant in the Kwa-Zulu Natal Province, and chrome mining is conducted in the Limpopo Province (Action Aid, 2019).

Interestingly, the rise of artisanal mining in the Limpopo Province coincided with retrenchments in the large-scale chrome mining industry in 2016. In 2016 alone, ASA Metals and Dilokong Chrome Mine retrenched 750 workers as part of restructuring plans. In total, it was reported that between 2015 and 2016, the chrome industry in South Africa shed more than 3000 jobs (Minerals Council, 2019). Furthermore, gold and platinum mines in the Limpopo Province also started stating their retrenchment plans in 2017. To illustrate the scenario, the leading platinum mine in the region, Bokoni Platinum, retrenched 2651 of its employees in 2017 (NUM 2021). The area is currently riddled with artisanal mining activities.

Falling Short: Policing Artisanal Mining

In 2023, President Cyril Ramaphosa authorized the deployment of 3300 South African National Defence Force (SANDF) personnel to support the South African Police Service in its ongoing operations against illegal mining for a 6-month period (State of the Nation, 2023). In the president's speech, it was correctly outlined that recent events have highlighted the severe impact of illegal mining on community safety. Notable incidents include the gang rapes of several women in 2022, allegedly by illegal miners; a gas explosion in May 2023 at an abandoned mine in Welkom that resulted in the deaths of 31 illegal miners; and another gas explosion in June 2023 at an informal settlement in Boksburg, which tragically took dozens of lives. These incidences culminated in the nationwide crackdown on illegal mining dubbed *Vala Umgodi*.

However, the issue of artisanal mining, globally and in the South African context, will not disappear. Given the illegal nature of the sector and the related social issues, including violence and conflict, it is understandable why all Zama Zamas are perceived as criminals. Unfortunately, this perception has led to a nearly exclusive focus on the illegality of Zama Zama mining. Attention is mainly directed toward resolving the issues related to artisanal mining and eradicating Zama Zama mining through law enforcement and policing. However, policing of the artisanal mining sector has seldom, if ever, been successful in South Africa or elsewhere (Bester & Uys, 2023; Hilson, 2017).

For example, in 2019, authorities in Brazil tasked the military with removing miners from the Amazon forest, which has been devastated by mining. Thousands of soldiers were deployed to the forest under the codename *Operation Green Brazil*. Following the arrest of many artisanal miners a few years later, environmentalists and some state officials admitted that the military could not end informal mining in Brazil. Likewise, in 2019, Peruvian officials launched *Operation Mercury* to eliminate informal gold mining in Madre de Dios, known as the epicenter of informal gold mining (Bansah et al., 2022). At least 1500 military and police officers evicted thousands of informal gold miners while destroying mining equipment, and military bases were established to prevent mining and other criminal activities such as human trafficking, which are allegedly linked to informal mining. Despite spending so much on military operations, informal artisanal mining remains an environmental threat in Peru (Bansah et al., 2022). Similarly, after several fatalities and injuries to both soldiers and miners, as well as the destruction of mining equipment, military forces in Guyana, Colombia, Bolivia, Ecuador, and Venezuela failed to put an end to informal mining (Bansah et al., 2022).

Hilson (2017) explains that policing of the sector in Ghana has failed because the focus was on using:

> … military intervention to … 'sanitise' the sector, a strategy that has never worked in Ghana or elsewhere because it focuses on the 'expressions' of the sector's informality—child labor, the infiltration of foreigners or in this case, environmental degradation—rather than addressing what has been responsible for creating its problems in the first place.

Bansah et al. (2022) describe that in 2017, the government enlisted 400 police and military personnel to form *Operation Vanguard*, which, under the authority of the Inter-Ministerial Committee on Illegal Mining (IMCIM), arrested and prosecuted informal miners and seized and destroyed much of their mining equipment. Individuals and academics have criticized this military strategy, which resulted in human rights violations (Bansah et al., 2022; Hilson, 2017). Also, despite these military efforts, illegal mining in Ghana is still widespread.

Instead, as Hilson (2017) suggests, the focus should be on identifying the drivers of the sector and developing ways to address the problems responsible for creating the sector and its negative impacts. Similarly, Martin (2019) argues that current law enforcement and policy responses in South Africa need to consider the poverty and socioeconomic drivers behind Zama Zama mining in establishing a viable artisanal mining sector. As Hilson (2017) points out, this begins by asking questions about why and how artisanal mining occurs. Only by gaining a more nuanced understanding of Zama Zama mining can one start exploring ways of addressing the sector more sustainably.

Nowhere has this sector been driven out, cleaned out, or sterilized successfully. On the contrary, it has been shown that every time the government or the private sector responds by policing it, it drives the sector underground and exacerbates the social ills. If policing the sector is failing, other ways need to be established to respond more responsibly to the sector. By examining the factors driving artisanal mining and identifying the responsibilities of mining companies in addressing these issues, we could reconsider the marginalization of Zama Zama mining. This shift might lead to replacing the current reactive strategies of policing and criminalization with more sustainable and proactive solutions to tackle the sector's challenges. The upcoming section examines the drivers, as identified in research by Bester and Uys (2023), that contribute to artisanal mining within the South African context.

Adding Fuel to the Fire: The Ongoing Unsustainability of Zama Zama Mining

Unemployment and Poverty

During the interviews, it was found that the high unemployment rate in South Africa is a major driving force behind Zama Zama mining. Artisanal mining is common in areas where formal employment opportunities are scarce (Mujere & Isidro, 2016). Interviewees emphasized that the country's unemployment and high poverty levels fuel Zama Zama mining. An interviewee added the following:

> "Well, we have a country where the real unemployment is approximately 40 percent of the population. We have 60 percent of people in South Africa earning less than R2500 a month. We have the movement that wants to put a minimum wage now at R3000 per month, which is wholly inadequate. We have the greatest inequality gap in terms of the Gini coefficient. Okay. So, people are trying to find means to support and survive."—Nicky, civil society

During a discussion with a mine manager from a mining company, he agreed that unemployment and poverty were critical drivers of Zama Zama mining:

> "So now they just go the Zama Zama route. And I mean with the current situation that we are in jobs are scarce. So, people are turning to this to get money."—Will, mine manager

These perceptions underscore mainstream research around why people engage in artisanal mining. Hilson's research (2016) found that alluvial gold panning had become a full-time economic activity in some regions of Zimbabwe, directly and indirectly supporting the livelihoods of over 500,000 Zimbabweans. Furthermore, Hilson (2016) explains that ASM in sub-Saharan Africa should be viewed as a means of subsistence rather than an entrepreneurial activity.

Given sub-Saharan Africa's deteriorating economic state and the ever-widening gap between the rich and the poor, ASM has been an economic lifesaver in some ways, providing countless numbers of jobs and supporting the livelihoods of millions of people in a landscape where job creation has proven elusive (Hilson, 2016). Similarly, when I probed a key informant during a conversation on what she thinks is creating the Zama Zama situation, she replied as follows:

> "Poverty. Unemployment. No jobs, no job creation in South Africa, it's just nothing there."—Patricia, key informant

Keeping the responsibilities of mining companies in mind, it is worthwhile to remember that in South Africa, the Broad-Based Socioeconomic Empowerment Charter of 2018 and the Mineral and Petroleum Resources Development Act (MPRDA) of 2002 emphasize mining companies' role and responsibility in job creation.

The Broad-Based Socioeconomic Empowerment Charter of 2018, or Mining Charter III, outlines the government's approach to mining industry regulation, stating that noncompliance will result in legal action. The Mining Charter III addresses ownership and employee transformation targets, aiming to promote equitable access to South Africa's mineral resources for all. Furthermore, it aims to increase opportunities for historically disadvantaged South Africans (HDSAs) to enter the mining and mineral industry and benefit from the country's mineral resources, thereby encouraging skills development, empowering HDSAs, and advancing the social and economic well-being of mining communities (Moraka & Van Rensburg, 2015). Ownership; Housing and Living Conditions; Employment Equity; Human Resource Development, Inclusive Procurement and Supplier Development, and Mine Community Development are all elements of the Mining Charter III.

Of course, mining companies cannot be the only ones to be blamed for the country's unemployment rate. Nonetheless, the often deplorable conditions in mining communities and the legislative requirements make mining companies responsible for creating jobs.

The following statement refers to the level of desperation and the fact that both the government and mining companies are responsible for creating employment. Interestingly, during a conversation with a civil society worker, the person considered mining companies to have an equal obligation to the government and stated that both sectors are failing to fulfill their responsibilities:

> "…with people that have worked in the mines that were not adequately compensated, retrenched or pensioned or their health dealt with, and are so desperate and have no other way of putting food on the table, which is squarely at the feet of government and the mining industry."—Nicky, civil society

Closely linked to this issue is that limited livelihood strategies exist in and around mining communities, which many interviewees argued was fueling Zama Zama mining.

Limited Livelihood Strategies

From a regulatory point of view, Mining Charter III requires mining companies to develop livelihood opportunities within mining communities as part of local economic and enterprise development elements. These opportunities must be sound economic enterprise development initiatives that promote community growth. However, during the interviews, it was discovered that few alternative livelihood options are available to communities, which inevitably drives Zama Zama mining in South Africa.

The following statements demonstrate many participants' reasoning on the issue of limited livelihood strategies. A Zama Zama stated that he participated in Zama Zama mining, because he had no other choice, despite the dangers.

> "This is a high-risk business. If anything comes to my side, I can stop going underground. I can stop working underground."—Tommy, Zama Zama

On another occasion with one of the key informants, it was also mentioned that having limited livelihood options is driving people to engage in Zama Zama mining by stating:

> "See the tragedy of poverty is the poor have very limited options. That's called the poverty trap, so they are forced, for example, to degrade their natural resource base or engage in artisanal mining because they have no other viable choices."—Mark, key informant

The mining industry's role in developing alternative livelihood strategies is supported by legislative requirements. The MPRDA and the Mining Charter III mandate that mining companies address socioeconomic development and sustainability in and around mining communities. Limited livelihood strategies within mining communities highlight the critical need for more robust efforts to address the issue.

Yet, mining companies are frequently accused of failing their mining communities. Action Aid's research in South Africa confirms this by stating that communities near mines suffer more negative impacts due to their closeness to the mines and the political, economic, and social barriers they face in holding corporations and the government accountable. Many local economic development projects fail, and communities rarely benefit from formal mining activities in an area (Action Aid, 2018).

Retrenchments

The ongoing retrenchments in the formal mining industry are a significant driver of Zama Zama mining in South Africa. Since 2012, employment in the mining industry has been declining. More than 73,000 jobs were lost between 2012 and 2020, according to the Minerals Council South Africa (2019, 2020, 2021). Furthermore, according to Stats SA, the mining sector saw a reduction of 6926 jobs, bringing the total employment down to 472,153 in the second quarter of 2024. This decline follows the loss of 2696 jobs in the first quarter of 2024, indicating a significant total decrease of 9622 jobs in 2024 (StatsSA, 2024).

Interviewees indicated their concern about the ongoing retrenchments in the formal mining sector and stated that mine retrenchments play a massive role in propelling artisanal mining. A mine manager had the following to say in this regard:

> "So, I can't talk about other shafts. But the shafts that I am at, we have definitely had people that we bring out has been ex-employees of the shaft. So, this shaft went through a restructuring process in, I think in 2012, it was before my time now, I can't recall the amount of people, but I think it was something close to a thousand people were retrenched. So those people, not a lot of them but some of them turned the Zama Zama route. So, they know the place, they get into contact with the Zamas, and they then go and operate underground cause they know the place."—Will, mine manager

During an interview with a member of civil society, the person reflected on the reasons why retrenched miners enter the artisanal mining sector:

> "Of course, there are many mine workers who have been left behind in Johannesburg [when] ones' mines have suddenly shut down. Very often, mines just shut down, and the retrenchment packages workers have aren't even enough to get them home to Mozambique or Zimbabwe or Lesotho or the Eastern Cape. So they just stay around on the mine property. And they themselves have got the expertise with which to actually find the mineral. And they start operating and mine on a small and survival scale."—Danny, civil society

Interviewees emphasized that the miners' underground knowledge and skill predispose retrenched miners to go the Zama Zama route. The retrenched miner has very few other options left to survive. Reliance on previous work experience and existing skills pushes people into the artisanal mining sector.

Studies conducted on artisanal mining in Zimbabwe discovered that nearly all miners cited harsh economic conditions resulting from worker retrenchment from paid employment and high unemployment levels as the main reasons for going into artisanal mining (Dreschler, 2001). Also, according to Hilson (2016), low agricultural yields caused by erratic rainfall have exacerbated the situation in Zimbabwe. For this reason, artisanal mining, particularly gold mining, is regarded as a cure-all for surviving in harsh living conditions.

Some researchers, such as Mkhize (2017), advocate for implementing initiatives to reduce the poverty levels of artisanal miners. Such initiatives should include training artisanal miners to improve their employability in the formal sector (Mkhize, 2017). Another suggestion is to train formal miners before they get retrenched. The social and labor plan (SLP)[2] requires formal mining companies in South Africa to have a downscaling and retrenchment process in place. It emphasizes that a mining company should address potential retrenchments through a skills development process. This introduces the idea of reskilling.

---

[2] To obtain and maintain a mining right, a social and labor plan must be submitted to the Department of Mineral Resources and Energy (DMRE). This plan is a legal requirement and must be renewed every 5 years for approval. The DMRE will not permit any mine to operate without an approved social and labor plan.

Failed Reskilling

During the interviews, it was found that not reskilling people sufficiently added to retrenched workers entering the Zama Zama mining sector. In South Africa, mining companies are currently addressing retrenchments through portable skills training, which they are legally required to provide to their employees; additionally, the mine's SLP requires that the retrenched be assisted in finding other suitable employment forms or sustainable livelihood opportunities.

However, interviewees examined this critically and explained that portable skills training is insufficient and appears to be failing in practice and, at least in some cases, fueling the current artisanal Zama Zama sector.

> "So, we teach this guy how to build, how to do electrical work, how to go and do car maintenance. How to be a mechanic, we teach him all other types of skills. So that when he gets retrenched, he actually can fall back onto something else. But the problem is, there is not a lot of money in that. Working on a mine for a guy that's got like a Standard six he is making more money than some people that's got degrees. Now we take him off we say you are not earning this R12,000 anymore. You must now go back to be an uneducated South African on the streets. What does he do he tries maybe to fix cars for a living, and he makes R4000 a month, and he sees he is not making money. Sometimes he doesn't make money. He can find a job in the private sector maybe as a builder what do you pay a builder, a minimum wage of R4000, R5000 a month. He is used to R12,000 a month. He is supporting a family of 20 people; he doesn't survive so what does he turn to. He turns to things that he knows. Now he has to do it illegal. He makes the R12,000, he even can make more than R12,000 being an illegal miner. So, he will turn to that side."—Will, mine manager

During a discussion, one interviewee stated that the approach to reskilling in the mining industry, as well as other industries, is flawed:

> "You know, our reskilling in general, not just in terms of the mines, yes but I think that the bigger part about an industry to do reskilling is that it should be finding ways to reskill in relation to its own corporate sectoral responsibilities. That's not to suggest that if a person who was a miner, you know, if the mine has, for example, some bursary scheme, which gives that person the opportunity to, I don't know, learn basic skills of administration … or bookkeeping or whatever it is, that white-collar work should be something that is not an option … or even plumbing or whatever. But the point is that, at least, the mining industries should start with their own responsibilities."—Jane, civil society

It is suggested that when reskilling takes place before retrenchments, the mining industry should link skills development to the demands of other economic sectors. Additionally, shifting from a retrenchment strategy to one focused on succession planning, along with upskilling and reskilling the workforce, could be beneficial. This approach allows a mining company to enhance its internal recruitment by utilizing succession planning as a workforce planning tool that aligns its employees with the company's business objective (Mining Industry Human Resources Council, 2013). According to research, succession planning increases company turnover by encouraging employee engagement and skill development (Mining Industry Human Resources Council, 2013).

Creating internal job opportunities before retrenchments is a more sustainable approach but may not always be possible. Therefore, it is important to acknowledge the value of portable skills training, but it seems that the lack of adequate portable skills training fails the retrenched employee. Creating an alternative path is critical, but such practices necessitate innovation, planning, and creativity.

During another discussion, a civil society worker emphasized that mining operations often require fluctuating labor forces—reducing staff during downturns and increasing them during booms. This cyclical reality necessitates careful planning.

> "The social and labor plan framework and MPRDA were structured along those lines that it was always known that mining by its nature is cyclical. So, you have to reduce your labor force from time to time and increase your labor force from time to time. So whenever there's a discussion around no-fault dismissals, those are some of the issues that need to be discussed around how much of the labor force you are cutting down and what you are going to do with the labor force in order to prepare them for a world out there where they don't have jobs. And the only thing that they know is mining, but what we see is that the emphasis is more on deciding on the packages, the financial packages. So someone is not doing their job as far as looking at the other avenues around what needs to be done in order to ensure that these people are integrated back into the community and integrated back into the mainstream labor force."—Themba, civil society

While financial packages are often prioritized, there is insufficient focus on helping those being retrenched transition into new roles or reintegrate into their communities. The implication is that merely providing financial compensation is inadequate; there needs to be a broader strategy to equip workers with skills for alternative employment opportunities. What is needed is a more holistic approach to labor management in the mining sector, advocating for proactive measures that prepare workers for a future beyond mining, rather than just addressing immediate financial concerns. This reflects a deeper social responsibility to ensure that workers are not left behind as the industry evolves.

Poor Mine Closure and Rehabilitation

According to a large body of knowledge, inadequate mine closure and rehabilitation create dreadful environments, resulting in various community issues: Mine waste, acid mine drainage (AMD), and soil erosion all outlast the lifespan of a mine (Krause & Snyman, 2014). Such negative legacies threaten community well-being and safety (Krause & Snyman, 2014).

Since 1903, South African laws have held mine owners responsible for mining impacts, which ended when they obtained a closure certificate. However, many mines became abandoned because of weak regulations. To address this, the 1975 Fanie Botha Accord established that the state would take full responsibility for mines closed before 1976, share responsibility for those closed between 1976 and 1986, and place all responsibility on owners for mines closed after 1986. Over the past decade, the Department of Water Affairs has spent over R120 million to address historic pollution, but this is only a small part of what is needed (Munnik, 2010).

Also, the requirements of the National Environmental and Management Act (NEMA) of 1998 introduced the concept of perpetual liability, assigning responsibility to the mining company even after a closure certificate is issued by the Minister responsible for mineral resources (Watson & Olalde, 2019). Despite these requirements, there remains a trend among mining companies to dodge their social and environmental responsibilities when it is time to close down.

It is no secret that where mines have closed, such as De Beers in Kleinzee, South Africa, communities have been left in dire straits with few job opportunities (see Cronjé et al., 2009).

Throughout some conversations, it was frequently stated that artisanal mining in South Africa is driven by insufficient planning and preparation for mine closure and rehabilitation. During one discussion with members from civil society organizations working in the mining community, it was discussed how mines were failing to fulfill their legal obligations, which inevitably drives Zama Zama activities:

> "… we do research on corporate social responsibility, and the major concern for us is the failure of closing mines properly. So, mines simply get abandoned. Okay. And when mines are abandoned that is when the possibility of small-scale mining opens up. So, what happens is when you have an abandoned mine site, usually what happens is it gets occupied by informal settle[ment]s because it is the only land available in Gauteng for occupation."—Danny, civil society

Another civil society worker mentioned that when mining companies merely walk away from concessions without addressing social and health issues in communities. This ultimately creates an environment in which Zama Zama mining flourishes and highly unsafe circumstances for community members.

> "What is important is the mines, why are we having Zama Zamas? It's because they will loot minerals, and run away without rehabilitating. These same shafts, are the very same shafts where we find our people. In Payneville we had Simpiwe who is 27 years old who fell into the shafts. As of today they couldn't find him. We have Richard the 5-year old boy who fell into these shafts last year in February, Richard in Boksburg. As of today, they couldn't find him. I can take you into that shaft now that shaft is not secure with no fence for people not to use that place. People live about 200 meters away, they pass there every day. Imagine how that mother of that boy feels, passing there every day knowing that my son is in there."—Mike, civil society

The ineffective closure of mines in South Africa is highlighted by the numerous abandoned mines, prolonged care-and-maintenance operations, and the transfer of mines to less financially capable companies, often leading to increased Zama Zama mining activities.

Despite these issues, there has been a lack of data to support these observations and understand the underlying problems. An analysis of closure certificates applied for and granted between 2011 and 2016 across all nine provinces in South Africa revealed interesting results. The analysis showed that the current mine closure system is largely ineffective. While closure certificates are being issued, they are mostly for prospecting sites and small-scale mines with minimal environmental impact. No significant large mines were closed during this period, and very few applied for closure certificates. Additionally, the issuance of closure certificates varies greatly between regional offices, with generally low success rates and lengthy processing times (Watson & Olalde, 2019).

Proper closure planning would lessen the impact on communities and a mining company's workforce while making room for the possibility of integrating the artisanal mining sector. On this note, a civil society worker said the following during a discussion:

> "Now, if you would begin to see your closure plans, as not just putting a cement block in a place where, you know, you don't actually need huge amounts of imagination or resources. But you start to see it as a much more holistic plan. You could then involve people who have no other form of employment in those kinds of operations."—Jane, civil society

Before mine closure, active efforts should be made to identify new projects to fill the void left when a mine closes. Preferably, members of the surrounding communities should be involved in developing such projects. Closure planning must be integrated into life-of-mine (LoM) planning and should consider environmental, social, and economic factors, as well as involve stakeholders and communities throughout the mining life cycle (ICMM, 2019). Projects may include ASM activities that can continue without formal mine management (Morrison-Saunders et al., 2015). Post-mine life considerations for ASM will likely require project-based analysis at the environmental impact assessment level (Morrison-Saunders et al., 2015).

The general oversight of ASM considerations in a relatively wealthy country, highlights the need for the private sector and governments to work together to solve problems associated with ASM (Morrison-Saunders et al., 2015).

However, the South African mining industry seems utterly unconcerned about developing the ASM sector. Furthermore, participants stated that Zama Zama mining is fueled by mining companies' casual attitude toward mine closure requirements. During an interview, a civil society worker said the following:

> "Well, it (the Zama Zama issue) is also the industry's responsibility in the sense that they have been making money out of South Africa for the last 130 years. Now at least, the least they can do as they move towards closure is to actually say we will assist the mine workers and the community that are here to become the small-scale miners in a meaningful way. In a way that is not as destructive as the current fashion is. Because now they are simply abandoning it and people are going into those mines without being properly prepared and train to do the work that they are doing."—Danny, civil society

This abandonment leaves people unprepared and untrained, leading to dangerous and unsustainable artisanal mining practices. The formal mining industry must take responsibility and support a transition to safer, regulated artisanal and small-scale mining.

Conclusion

When examining artisanal mining, it is essential to acknowledge that the issue is complex. It is a complicated sector layered by a multitude of factors and drivers. The country's high unemployment rate and the mine's corporate responsibility recognized in legislative and regulatory frameworks demand a different take on the current Zama Zama mining situation. Alternatively, at the very least, drivers such as the retrenchment process, and mine closure and rehabilitation need to be reconsidered and restructured. In terms of poor mine closure and rehabilitation, the MPRDA and NEMA require governing processes for mine closure and rehabilitation. However, it has been recorded that there are many abandoned mines in South Africa, ultimately fueling the current Zama Zama situation.

It is also imperative to acknowledge the rich diversity of the sector. As an analysis of the drivers has indicated, several socioeconomic factors are driving Zama Zama mining, of which a preference for criminality is often not one. To improve the situation, innovative strategies for handling retrenchments are necessary, such as re-employing affected workers internally, when this is possible. One participant suggested that allowing retrenched employees to engage in small-scale mining could extend the mine's operational life when the mine approaches closure. This approach would likely foster a more sustainable environment for artisanal mining, enabling it to become a meaningful economic contributor.

This chapter provided a comprehensive overview of the profile of the artisanal miners in South Africa. Part of the profile and a dimension often ignored within the sector is the role women play in this sector. For that reason, the next chapter provides a detailed look at the positionality of women within the artisanal mining sector and women Zama Zamas in the South African context.

References

Action Aid. (2018). Mining in South Africa 2018. Social Audit Baseline Report. Accessed from, November 21, 2023, from https://www.wits.ac.za/media/wits-university/faculties-and-schools/commerce-law-and-management/research-entities/cals/documents/programmes/environment/resources/Social%20audit%20report.pdf

Action Aid. (2019). *Regularising informal artisanal mining in South Africa: An evidence-based report*. Action Aid.

Bansah, K. J., Acquah, P. J., & Assan, E. (2022). Guns and fires: The use of military force to eradicate informal mining. *The Extractive Industries and Society, 11*, 101139. https://doi.org/10.1016/j.exis.2022.101139

Barreto, M. L., Schein, P., Hinton, J., Hruschka, F.. (2018). *The impact of small-scale mining operations on economies and livelihoods in low- to middle-income countries*. Pact Inc and Alliance for Responsible Mining. Accessed May 27, 2021, from https://www.gov.uk/research-for-development-outputs/the-impact-of-small-scale-mining-operations-on-economies-and-livelihoods-in-low-to-middle-income-countries

Bench Marks Foundation. (2017, March 29). *DMR called to legitimize "Zama-Zama" mining*. Press release. https://www.polity.org.za/article/bmf-dmr-called-to-legitimise-zama-zama-mining-2017-03-30

Bester, V. (2019). *A corporate social responsibility (CSR) conceptual framework to address unregulated artisanal gold mining in South Africa.* Doctoral Thesis / Master's Dissertation. Johannesburg: University of Johannesburg. Accessed November 3, 2024, from https://ujcontent.uj.ac.za/esploro/outputs/doctoral/A-corporate-social-responsibility-CSR-conceptual/9911598107691

Bester, V., & Uys, T. (2023). Artisanal mining and its drivers in the South African context. *The Extractive Industries and Society, 15*, 101278. https://doi.org/10.1016/j.exis.2023.101278

Buxton, A. (2013). *Responding to the challenge of artisanal and small-scale mining. How can knowledge networks help?* Accessed August 15, 2015, from http://pubs.iied.org/16532IIED.html

Cronjé, J. F., van Wyk, D., & van Wyk, L. J. (2009). *SADC research: Corporate social responsibility in the diamond mining industry on the west coast of South Africa*. Research Done on Behalf of the Netherlands Institute for Southern Africa (NIZA) and The Bench Marks Foundation. Johannesburg, South Africa. Accessed March 20, 2019, from https://www.bench-marks.org.za/policy-gap-4/

Collins, N., & Lawson, L. (2014). *Investigating approaches to working with artisanal and small-scale miners: A compendium of strategies and reports from the field*. Accessed July 5, 2018, from https://www.csrm.uq.edu.au/media/docs/1214/collins-asm-fr-completed-report.pdf

De Haan, J., Dales, K., & McQuilken, J. (2020). *Mapping artisanal and small-scale mining to the sustainable development goals. Minerals, Materials and Society program in partnership with Pact*. Accessed January 20, 2022, from https://www.pactworld.org/library/mapping-artisanal-and-small-scale-mining-sustainable-development-goals

Department of Mineral Resources, South Africa. (2017). 'Small Scale Mining'. Accessed 3 July 2019 from: https://www.dmr.gov.za/mineral-policy-promotion/small-scale-mining

Department of Mineral Resources and Energy. (2022). *Mineral and Petroleum Resources Development Act: Artisanal and small scale-mining policy 2022* | South African Government. https://www.gov.za/documents/mineral-and-petroleum-resources-development-act-artisanal-and-small-scale-mining-policy

Dreschler, B. (2001). *Small-scale mining and sustainable development within the SADC region*. Accessed May 20, 2023, from https://commdev.org/wp-content/uploads/pdf/publications/Small-Scale-Mining-Development-SADC-Region.pdf

ENCA. (2019, June 6). *Zama Zama cost SA R41bn in 2018*. ENCA. Accessed July 17, 2019, from https://www.enca.com/news/zama-zamas-cost-sa-r41bn-2018

Groenewald, L., Huchzermeyer, M., Kornienko, K., Tredoux, M., Rubin, M., & Raposo, I. M. (2013). Breaking down the binary: Meanings of informal settlement in southern African cities. In S. Bekker & L. Fourchard (Eds.), *Governing cities in Africa* (1st ed.). HSRC Press. https://scholar.sun.ac.za/bitstream/10019.1/90428/1/groenewald_breaking_2013.pdf

Hilson, G. (2009). Small-scale mining, poverty and economic development in sub-Saharan Africa: An overview. *Resources Policy, 34*(1), 1–5.

Hilson, G. (2016). Farming, small-scale mining and rural livelihoods in Sub-Saharan Africa: A critical overview. *The Extractive Industries and Society, 3*(2), 547–563. https://doi.org/10.1016/j.exis.2016.02.003

Hilson, G. (2017). Shootings and burning excavators: Some rapid reflections on the Government of Ghana's handling of the informal Galamsey mining 'menace'. *Resources Policy, 54*, 109–116. https://doi.org/10.1016/j.resourpol.2017.09.009

Hilson, G., & McQuilken, J. (2014). Four decades of support for artisanal and small-scale mining in sub-Saharan Africa: A critical review. *The Extractive Industries and Society, 1*(1), 104–118. https://doi.org/10.1016/j.exis.2014.01.002

Huchzermeyer, M., & Karam, A. (2006). *Informal settlements: A perpetual challenge?* UCT Press.

International Council on Mining and Metals, The. (2019). *Integrated mine closure: Good practice guide* (2nd ed.). Accessed May 20, 2023, from https://www.icmm.com/en-gb/guidance/environmental-stewardship/2019/integrated-mine-closure

International Institute for Sustainable Development. (2017). *Global trends in artisanal and small-scale mining (ASM): A review of key numbers and issues*. Accessed February 22, 2023, from https://www.iisd.org/system/files/publications/igf-asm-global-trends.pdf

International Labour Organization. (2019). *Child labour in mining and global supply chains.* https://www.ilo.org/wcmsp5/groups/public/%2D%2D-asia/%2D%2D-ro-bangkok/%2D%2D-ilo-manila/documents/publication/wcms_720743.pdf
Krause, R. D., & Snyman, L. G. (2014). *Rehabilitation and mine closure liability: An assessment of the accountability of the system to communities*. Accessed January 16, 2019, from https://www.wits.ac.za/media/wits-university/faculties-and-schools/commerce-law-and-management/research-entities/cals/documents/Rehabilitation%20and%20mine%20closure%20liability.pdf
Legal Resources Centre. (2016). *Artisanal Mining Report*. Accessed April 25, 2018, from http://lrc.org.za/wp-content/uploads/pdf/2016_Artisanal%20Mining%20Report_LRC%20Publication.pdf
Martin, A. (2019). *Uncovered: The dark world of the Zama Zamas*. ENACT Africa. Accessed October 30, 2024, from https://enactafrica.org/research/policy-briefs/uncovered-the-dark-world-of-the-zama-zamas
Minerals Council South Africa. (2019). *Facts and figures 2018*. https://www.mineralscouncil.org.za/industry-news/publications/facts-and-figures
Minerals Council South Africa. (2020). *Facts and figures 2019. Chamber of Mines*. Accessed November 20, 2021, from https://www.mineralscouncil.org.za/industry-news/publications/facts-and-figures
Minerals Council South Africa. (2021). *Facts and Figures 2020 Pocketbook*. https://www.mineralscouncil.org.za/industry-news/publications/facts-and-figures
Mining Industry Human Resources Council. (2013). *Managing through the cycle: A strategic approach to workforce planning in the mining industry*. Accessed March 1, 2021, from Managing-through-the-Cycle.pdf (mihr.ca)
Mkhize, M. (2017). New interventions and sustainable solutions. Reappraising illegal artisanal mining in South Africa. *SA Crime Quarterly*. Accessed December 15, 2018, from http://www.scielo.org.za/scielo.php?script=sci_arttextandpid=S1991-38772017000300007
Mujere, N., & Isidiro, M. (2016). *Impacts of artisanal and small-scale gold mining on water quality in Mozambique and Zimbabwe*. https://pdfs.semanticscholar.org/0ae3/4a80b260c8eb4c55b375bf972172155f8166.pdf.
Moraka, N. V., & Van Rensburg, M. J. (2015). Transformation in the South African mining industry – looking beyond the employment equity scorecard. *Journal of the South African Institute of Mining and Metallurgy, 115*(8), 669–678. https://doi.org/10.17159/2411-9717/2015/v115n8a2
Morrison-Saunders, A., McHenry, M. P., Wessels, J.-A., Sequeira, A. R., Mtegha, H., & Doepel, D. (2015). Planning for artisanal and small-scale mining during EIA: Exploring the potential. *The Extractive Industries and Society, 2*(4), 813–819. https://doi.org/10.1016/j.exis.2015.08.004
Munnik, V. (2010). *The social and environmental consequences of coal mining in South Africa: A case study*. Accessed March 20, 2020, from http://www.bothends.org/uploaded_files/uploadlibraryitem/1case_study_South_Africa_updated.pdf
Mutemeri, N., & Petersen, F. W. (2002). Small–scale mining in South Africa: Past, present and future. *Natural Resources Forum, 26*(4), 286–292. https://doi.org/10.1111/1477-8947.t01-1-00031
O'Faircheallaigh, C., & Corbett, T. (2016). Understanding and improving policy and regulatory responses to artisanal and small scale mining. *The Extractive Industries and Society, 3*(4), 961–971. https://doi.org/10.1016/j.exis.2016.11.002
Perks, R., & Schneck, N. (2021). COVID-19 in artisanal and small-scale mining communities: Preliminary results from a global rapid data collection exercise. *Environmental Science & Policy, 121*, 37–41. https://doi.org/10.1016/j.envsci.2021.03.007
South African Human Rights Commission (SAHRC). (2015). *Report of the SAHRC investigative hearing issues and challenges in relation to unregulated artisanal underground and surface mining activities in South Africa*. Accessed November 12, 2015, from https://www.sahrc.org.za/home/21/files/Unregulated%20Artisanal%20Underground%20and%20Surface%20Mining%20Activities%20electronic%20version.pdf
South African Government. (2003). National Small Business Amendment Act 26 of 2003 | South African Government. Accessed November 17, 2021, from www.gov.za

State of the Nation. (2023, November 13). *Together we will win the battle against illegal mining*. Accessed November 17, 2024, from https://www.stateofthenation.gov.za/newsletter/together-we-will-win-the-battle-against-illegal-mining

StatsSA. (2024). *Quarterly Labour Force (QLFS) Q2: 2024*. Accessed October 1, 2024, from https://www.statssa.gov.za/publications/P0211/Presentation%20QLFS%20Q2%202024.pdf

The Borgen Project. (n.d.). *10 quotes on poverty*. Accessed November 15, 2024, from https://borgenproject.org/10-quotes-poverty/

The University of British Columbia and Canadian International Resources and Development Institute (CIRDI). (2019). *Artisanal and small-scale mining: Mapping the South African ecosystem*. Accessed November 19, 2021, from https://cirdi.ca/artisanal-and-small-scale-mining-mapping-the-south-african-ecosystem/

Toteu, S. (2017, November 28). *Transforming artisanal and small-scale mining in Africa through research and training*. Accessed May 8, 2017, from https://www.assaf.org.za/files/Science%20BUsiness%20Dialogue%202017/Transforming%20Artisanal%20and%20Small-Scale%20Mining%20in%20Africa%20Felix%20Toteu.pdf

United Nations Economic Commission for Africa. (2011). *Minerals and Africa's development*. Accessed November 19, 2021, from https://archive.uneca.org/publications/minerals-and-africas-development

Watson, I., & Olalde, M. (2019). The state of mine closure in South Africa – what the numbers say. *Journal of the Southern African Institute of Mining and Metallurgy, 119*(7), 639–645. https://doi.org/10.17159/2411-9717/331/2019

World Bank. (2020). *2020 State of the Artisanal and small-scale mining sector*. Accessed November 19, 2021, from https://documents1.worldbank.org/curated/en/884541630559615834/pdf/Delve-2020-State-of-the-Artisanal-and-Small-Scale-Mining-Sector.pdf

World Bank. (2024). *Achieving sustainable and inclusive artisanal and small-scale mining (ASM): A renewed framework for world bank engagements*. Accessed September 28, 2024, from https://openknowledge.worldbank.org/server/api/core/bitstreams/cab95ce4-781f-42df-9c4d-912c55834c1b/content

Chapter 5
Delving Deeper into the World of Women in Artisanal Mining

I raise up my voice—not so that I can shout, but so that those without a voice can be heard. We cannot all succeed when half of us are held back.

—Malala Yousafzai (2013)

Introduction

It has been reported that women make up 30% of the global artisanal and small-scale gold mining (ASGM) workforce. Furthermore, the World Health Organisation (2016) reported that women comprise 10% of Asia's artisanal and small-scale mining (ASM) population, 20% in Latin America, and 40–50% in Africa.

Women's experiences in this sector vary from men's for a number of reasons, which will be explored in this chapter. Also, conversations with Zama Zama women and other key informants highlight the rich diversity of women's experiences in this field. These differences and nuances should be considered when researching the matter and when developmental initiatives are developed to support the formalization process. This chapter gives an overview of the overall challenges women face in the sector, ranging from the division of labor to gender norms, restricted access to capital and resources, and specific health and safety issues, including the issue of sexual harassment. Yet, simultaneously, some women see artisanal mining as a viable livelihood and as a means of economic empowerment.

Women Artisanal Miners and International Policy

Many researchers report that development agencies and donors have overlooked women's role in ASM. This is a somewhat misplaced perception. Women's participation in ASM is a topic that has not gone unnoticed; the discussion already began in the 1990s (Hilson et al., 2018). Hilson et al. (2018) explain that Labonne (1996) produced one of the earliest and most critical pieces, emphasizing the importance of acknowledging the women themselves and prescribing a series of steps that were presumably intended for policymakers (Hilson et al., 2018). Around the same time,

V. Bester, *The Untold Story of Zama Zama Miners in South Africa*, Clinical Sociology: Research and Practice, https://doi.org/10.1007/978-3-031-82783-9_5

the United Nations published a comprehensive report on the ASM sector (see UN, 1996). The report states that *"women constitute a significant portion of the human resources in small-scale mining and artisanal mining, and in some areas, they are the predominant participants."* The report also recognizes women's diverse roles at ASM sites, such as concession owners, hawkers and providers of services and goods, workers, dealers and buying agents, food preparers, and equipment owners who hire shovels and pans. Other reports (for example, Davidson, 1993; Traore, 1994; Burke, 1997) highlight similar points about women in ASM, many of which are relevant to sub-Saharan Africa (Hilson et al., 2018).

However, because the recommendations were made in a vacuum, this research body had no impact on the policymaking and donor space at the time. In the 1990s, organizations such as the World Bank, the United Nations Department of Economic and Social Affairs, the Intermediate Technology and Development Group, and the German Development Agency (GTZ) funded hopeful technical, policy, and geological work for ASM. However, much of it was siloed and disconnected from broader development policies and dialogues at the time (Hilson et al., 2018). Failure to incorporate ASM into mainstream development literature in the 1990s and early 2000s is a significant reason why the struggles and experiences of the women in artisanal mining were virtually ignored, even though gender in sub-Saharan Africa and the experiences of the region's women entrepreneurs were attracting considerable attention in the academic community at the time (Hilson et al., 2018).

This oversight by donors and policymakers calls for a more critical analysis of gender power relations that structure women's involvement in ASM (Danielsen & Hinton, 2020). What further added to the oversight of women in the ASM sector is the dominant perception that mining is a male-only activity despite the increasing participation of women in the sector (Arthur-Holmes & Busia, 2022). This perception is still dominant in both the formal and informal mining sectors.

Women are seldom recognized as miners in the artisanal mining space, although they partake in digging, sluicing, separating gold, and burning amalgam. Women's participation in mining activities is currently under-recognized and under-theorized.

It is critical to recognize women in this sector as part of the formal mining sector's role concerning development and poor communities in the global South (Jenkins, 2014). According to Traore (cited in Hinton et al., 2003), *"the positive effects [of artisanal mining] are hardly felt by women, [but] they are hard-hit by the negative effects."* This is due to women's unequal position in their families and communities, and artisanal mining and its consequences exacerbate these inequalities.

Women often face obstacles in accessing capital, such as needing approval from their husbands or male relatives and providing collateral, especially land (Danielsen and Hinton (2020). For example, women in the Great Lakes Region (GLR) struggle to control and own land due to biased inheritance laws, needing permission from husbands or male relatives to buy land, a lack of capital, and limited influence over land decisions (Perks, Kelly, Constantian and Pham, 2018).

The same issue exists in South Africa, where women struggle to access land and mineral rights, with no support currently available for female miners to obtain mineral rights licenses (Action Aid, 2019). Supportive of this, Lahiri-Dutt (2015) argues that informal mining is still under-recognized in the mining literature; it remains a complex terrain due to the complex issues of property rights that contribute to the illegitimacy of this livelihood. Global economic policies often push women into low-wage, insecure, and risky jobs in informal mining. Often, they do not own land or capital, and their low social and economic status makes them the most vulnerable workers in this sector (Beneria, 2010). As a result, global policymakers generally remain silent on gender issues in informal mining (World Bank, 2024). Consequently, researching women's challenges in ASM is critical for addressing their persistent *"stay-behind"* roles, which increase their economic marginalization and socio-cultural invisibility in the sector, as well as advocating for gender mainstreaming in ASM (Arthur-Holmes & Busia, 2022).

While ASM researchers, activists, and practitioners use and apply the concept of gender in various ways, Hinton discovered that one shared understanding is that women and men have different, socially ascribed productive, reproductive, and social roles that give rise to different needs. This understanding of gender differences highlights the diverse roles of women and men in ASM within their households, communities, and mines.

Therefore, before we continue with this chapter, it is essential to define what is meant by gender. For this chapter and the book, gender is considered as follows: Unless otherwise stated, gender refers explicitly to the behavior, attitudes, values, and beliefs a specific sociocultural group considers appropriate for males and females. Gender roles, it is frequently suggested, are fluid and can shift over time, space, and context (Hinton et al., 2003). Gender roles refer to behaviors, tasks, and responsibilities assigned to women and men in a given society because they are considered appropriate (Hinton, 2016).

Regardless of socioeconomic class, systematic gender differences in material well-being can exist, though the degree of inequality varies across countries and over time (Meinzen-Dick et al., 2014). It is also vital to acknowledge that gender inequality is widespread in most societies, with males holding higher positions in social, economic, and political hierarchies (Meinzen-Dick et al., 2014).

The Role of Women in Artisanal Mining

Concerning women's direct involvement in ASM, scholars have noted that women's roles, determined by socio-cultural norms, include the provision of water on mineralized sand, transportation of mineral ores, and serving as forewomen (Arthur-Holmes & Busia, 2022; Danielsen & Hinton, 2020). The literature also indicates that local material transportation, for example, is almost always done by women in head-loads of baskets. In contrast, technical jobs requiring skill or the use of machines are almost always reserved for men. In addition, indirectly, women play roles such as cooks, nightclub entertainers, sex workers, and merchants (Ofosu et al., 2022).

According to some artisanal mine owners, women are docile, with *"nimble fingers,"* and are not supposed to do heavy work (Lahiri-Dutt, 2008). Furthermore, Lahiri-Dutt (2015) explains that the proportion of women working on artisanal mining sites varies according to location, mineral nature and value, processing techniques, marketing systems, and local social circumstances. Lahiri-Dutt (2015) clarifies that it is essential to note that the percentage of women in mining jobs can vary; for example, it may range from 10% to 60% at a particular site (Hinton et al., 2006; Olayide et al., 2013). However, despite the higher number of women artisanal miners at a site, masculinity remains the predominant cultural value (Cuvelier, 2014). Also, even though many women work long hours in mining, doing work as tedious as men, they rarely self-identify or are identified by others as miners. *"The man is the miner syndrome, discriminates against women's access to a variety of miners' resources and mineral rights"* (Danielsen & Hinton, 2020). Consequently, women are frequently recognized as supplementary workers, working in the sector temporarily (Ofosu et al., 2022).

It has been observed that the dangerous and physically demanding work of ASM leads to the gender division of labor in which men undertake the roles requiring manual lifting of heavy loads while women undertake repetitive chores.

Another significant labor aspect is that women are sometimes excluded from mine shafts altogether for taboo-related reasons. In some communities, it is thought that menstrual blood or traces of sexual activity on women's bodies may *"chase"* away mineral ores, especially gold (Ofosu et al., 2022). Also, in Arthur–Holmes and Busia's research (2022) set in Ghana, cultural norms, superstitious beliefs, and social taboos encouraged this gendered order, which translated into women miners taking lower-level supporting roles compared to managerial roles and, ultimately, ownership of mine sites. Another example is that in many countries, it is believed that a woman's presence in mines is thought to cause minerals to *"disappear,"* and if a woman whistles near a mining area, an accident will occur, among other things (Danielsen & Hinton, 2020). These cultural taboos contribute to women's exclusion from artisanal mining sites.

Also, Danielsen and Hinton's research (2020) indicates that some men appear to believe that in this male-dominated environment, women are more likely to engage in sex with co-workers. "*It is not easy to convince your husband that you want to join a group of about six men when you are one woman,*" said a female tin miner (Danielsen & Hinton, 2020). Most men do not believe a woman can work with men "*(without) engaging in any sexual relations*" (Danielsen & Hinton, 2020).

Indeed, several women use sex as a coping mechanism. In one Ugandan tin mine, women traded sexual favors for men's assistance in breaking tough rock and cases of women trading sex for mineralized sand have been described (Hinton & Mbabazi, 2009).

In addition, research discovered that 18% of women in the Kivu region of the Democratic Republic of the Congo (DRC) had traded sex for access to work in mines, cafes, and other jobs at some point (Perks et al., 2018). Notably, Perks et al. (2018) discovered that single and married women were less likely to engage in sex work than widowed or divorced women, implying that sex is used as a coping strategy by women who face economic and social exclusion (Danielsen & Hinton, 2020).

Women occupy marginal roles in managing ASM operations (Arthur-Holmes & Busia, 2022). They rarely attain the same decision-making positions as their male counterparts, including concession owners and mine operators (Ofosu et al., 2022). Jobs in small mines and quarries are gender segregated, reflecting what is commonly called horizontal segregation, with women having restricted access to specific jobs.

Research on artisanal mining in South Africa supports this by saying that there is no gender equality in the sector, and female miners are discriminated against in terms of where they can work on the artisanal mining sites and the resources that they exploit (Action Aid, 2019). The working relationships and arrangements on sites favor male counterparts. In addition, the research shows that women rarely work underground; they only participate in ore extraction when deposits have thin overburden layers (Malpeli & Chirico, 2013). Mineral deposits' geologic and geomorphic expressions primarily determine their location and accessibility, and the required types of effort and skill influence women's roles in ASM. For instance, women are primarily assigned to other roles when deposits require significant manual labor to access ores because of thick, overburdened layers (Malpeli & Chirico, 2013).

Another aspect that feeds into the gender-based division of labor is women's household obligations. These obligations impact women's livelihoods by reducing their time spent doing mining work. Buss et al. (2021) explain that women's social reproductive roles are not only household commitments but also: "*the activities and attitudes, behaviors and emotions, responsibilities and relationships directly involved in the maintenance of life on a daily basis, and intergenerationally...*" These roles shape the types of mining work women are seen as suited for while downplaying the importance of women's mining livelihoods.

Despite the high prevalence of women in the artisanal mining sector, there is a common misconception, particularly in South Africa, that only men are involved in Zama Zama mining.

I observed many women participating in artisanal mining during visits to various Zama Zama sites. Although exact figures are unavailable, fieldwork observations and interviews confirm that many women are part of the sector. Interviewees noted that women rarely go underground, though some do, which aligns with existing literature on gender divisions. Two key informants elaborated on the role and status of women as follows:

> "They only do the processing. I've never come across a woman who has gone underground."—Carina, key informant

> "Women do have a role but that role, I think is right at the bottom level of production and probably doesn't bring them very much, by way of income and so on. I don't think they get significant livelihoods from it. You've got the processing, which is a mixture of men and women. You often have like a support system that's there as well, for the illegal mining, so women bringing food and so on."—Teo, key informant.

During an interview with a Zama Zama male miner, the role of women was confirmed. He said the following:

> "Yes, we are working with them (women), but they don't go underground. They do the crushing."—Stuart, Zama Zama

Figures 5.1 and 5.2 show women involved in the processing part. Both photos were taken during the field research.

During an interview with a key informant, he elaborated on the division of labor in artisanal mining by explaining the intricate structure of the mining sector, particularly Zama Zama mining. During the conversation, it was emphasized that while men primarily handle the extraction, often involving underground work, women and children are more involved in surface operations, like crushing and processing. He explained that in countries like Senegal and Mali, women also play a significant role in gold extraction, even though they face considerable risks, such as exposure to mercury. In this context, the interviewee explained that women often occupy supportive roles, providing food and other necessities to the miners. However, they generally remain at the lower end of the production hierarchy, which limits their economic benefits. The middlemen, predominantly men, manage the operations and syndicates purchase the minerals, indicating a gendered division of labor and power within this informal mining economy. This reflects a broader pattern of social and

Fig. 5.1 A woman Zama Zama turning a penduka (Bester, 2019)

Fig. 5.2 A woman artisanal miner (Bester, 2023. Unpublished)

economic exclusion, with women often receiving minimal income and recognition for their labor.

> "As you'll know, it's a whole complex production sector, where fairly typically, you would have young men, particularly if they go underground, doing the mining. In some cases, it's surface resources as well, in which case women and children may well be involved. Women fairly typically are involved in crushing rock and in processing. I know in other countries, in Senegal, for example and Mali, women are involved in extracting the gold using mercury. That's right and are exposed to those risks as well. There's the extraction, which is mostly young men, as I understand it and it will vary, of course. You've got the processing, which is a mixture of men and women. You often have like a support system that's there as well, for the illegal mining, so women bringing food and so on. Then of course, you have the

middlemen, who are the local managers, who I think are also, probably mostly men. Then of course, you have the syndicates that buy the minerals from them."—Teo, key informant

Surprisingly, the interviews and site visits I conducted do not fully support the previously mentioned views regarding women's roles in the sector. Data on women Zama Zamas was gathered at a well-established Zama Zama mining site in Roodepoort, Gauteng Province. During a group discussion with key informants, it was explained that women sometimes go underground and that the gendered division of labor will depend on the Zama Zama mining site. Consider the following exchange:

Researcher: "Do you know of many women that go underground?"
Patricia, Key informant: "Not many, but we know of and quite a number of them."
Ross, Key informant: "Yeah."
Researcher: "Really?"
Ross, key informant: "Yeah."
Researcher: "And they, but they're not at Matholeville?"
Patricia, key informant: "They are, there are those that are going underground actually at Matholeville, there are yes, there are that are going underground."
Researcher: "The men allow them to go underground?"
Tess, key informant: "Yes."

Some discussions highlighted a mutually beneficial relationship between men and women at Zama Zama sites. One Zama Zama woman explained that she owned a *penduka*[1] and informally employed the men to go underground. She explained the situation as follows:

"Okay, so there are men going underground that I sponsor with the money. Because there where by they mine, there is the security. They need pay money to go underground. They need food, they need water. Everything, they need cigarettes. So I pay for them. When they come back, they went to my base, my *penduka*. When they are done the work they leave the sand, the secondhand things. They take the first money and stay so we take the second money."—Esha, woman Zama Zama

It was clarified that the men Zama Zamas take the first grade (or round) of gold, and then they continue to crush, and Esha then takes the second and third grade of ore, which she then gets processed and sells.

Health and Safety Issues

Health and safety issues are rife in ASM and receive very little attention at most ASM sites, as precautionary measures are neglected, and no training on health and safety issues is provided (Ofosu et al., 2022). Furthermore, women are also exposed to health and safety risks because of limited information, such as the chemical hazards from mineral processing and how to prevent exposure (Action Aid, 2019).

[1] A *penduka* is a cylinder, often an old gas bottle, with iron balls that is turned for hours to crush the ore inside.

Other possible health and safety risks women face are as follows: brain, kidney, and lung damage because it is often women who perform ore purification using mercury or cyanide and scavenge tailings where they come into contact with chemicals such as cyanide (IIED, 2018). Moreover, women are also more vulnerable to health problems such as lung disease, skin irritation, eye damage caused by ore-crushing dust, and musculoskeletal disorders caused by manual lifting and carrying of materials and awkward positioning.

Women are also drained due to the physical demands of performing the above-mentioned activities and the domestic responsibilities they are traditionally assigned, such as childcare, food and drink preparation, and cleaning, among other things. The cultural constraints are more profound, as women are often at the bottom of the family hierarchy regarding rest, comfort, and nutrition in the home, making them more vulnerable to health risks (IIED, 2018).

For women, the gender division of labor poses a specific set of health and safety challenges. The gender segregation of mine work frequently explains the different activities and tasks that men and women perform in the sector. This segregation affects their occupational health and safety. Compared to large-scale mines that use modern mining and extraction methods, most ASM sites use rudimentary equipment and basic methods; this likely increases women's occupational exposure to physical and chemical hazards (Arthur-Holmes & Busia, 2022).

When I asked about the health and safety challenges that women Zama Zamas face in general, a key informant elaborated and confirmed that women face, in some instances, disturbing challenges.

> "That lady like, they have got this problem with their men, or those men, they are working with the men, they oppress them because they are women, and believe that women shouldn't do, and that women don't belong to that particular area. As men, they have got this power over the women and they even take their stuff and firstly, they don't take like nicely. … they come and take their money or their stuff sometimes."

> Researcher: "So it's very site-specific."
>
> Ross: "Yes, in Krugersdorp, mostly because I know the ladies from Krugersdorp because they are always complaining about that."

The matter of these women's children also came to the fore. Through the field observations, I observed many children alongside their mothers. The babies are exposed to unhealthy environments. For example, at one Zama Zama site (see Fig. 5.3), a toddler stood beside where the mercury was mixed and occasionally stuck his hand in the mixture. This is just one example of the danger children face at the Zama Zama site.

Carina, a key informant, said the following about women and their children:

> "…and very often a lot of them have got their children with them. Babies on their backs with that dust…"

A Zama Zama woman confirmed this by saying the following when I queried about what she does with her baby when she has to work:

> Researcher: "What do you do with the baby when you work?"

Fig. 5.3 A toddler next to where mercury gets mixed (Bester, 2019)

> Zama Zama woman: "I keep the baby at my back, I work with the baby at my back."

When the gendered structures that influence gender roles and decision-making at mine sites are changed, women's health and safety needs in informal workplaces like ASM stand to benefit. In this way, increasing women's involvement and presence in management and decision-level ranks would be an excellent first step toward ensuring gender-sensitive policies and measures that reflect women's underserved health and safety concerns in ASM. This is especially important given the gender differences in risk exposure and the varying social contexts that result in gender-specific occupational health issues (Arthur-Holmes & Busia, 2022).

Sexual Harassment and Exploitation

Women in artisanal mining societies are commonly subject to violence, which has been documented on a global scale (Environmental Law Institute, 2014). For example, Bafilemba and Lezhnev (2015) stated that Sexual and Gender-Based Violence (SGBV) is common in mining areas in eastern DRC. The United Nations has documented high rates of SGBV by armed groups and army commanders in some countries where artisanal mining occurs. For example, according to the United Nations Security Council (2015) "*...sexual violence, including forced marriage, continues to be a problem in the eastern Democratic Republic of the Congo,*" a region well known for its artisanal mining.

It was reported that "*...during the first six months of 2014, the United Nations Population Fund documented 2,774 cases of sexual violence only in North Kivu*" of the DRC (United Nations Security Council, 2015).

These incidents are likely to cause a spike in Human Immunodeficiency Virus/ Acquired Immunodeficiency Syndrome (HIV/AIDS) and other sexually transmitted diseases in a community, having an undeniably negative effect on the community (Environmental Law Institute, 2014). Without a doubt, women artisanal miners face disproportionate challenges in accessing lucrative ore deposits, mining groups, equipment, finance, and fair markets, with increased sexual and gender-based violence (SGBV).

The risks women face as part of the Zama Zama mining sector are worrying. During my interviews, it was pointed out that women are in a more subordinate position to men in the Zama Zama sector. In other words, in an already marginalized group, women are further suppressed and marginalized by the actors in the groups and those that are outside of the sector. Women artisanal miners are discriminated against by their male counterparts, who consider them too weak to manage (Action Aid, 2019). During an interview, it was explained that women's role is primarily about supporting men and it was detailed how they are getting exploited. The following quotations shed some light on the aforesaid:

> "They are exploited in every possible way, especially the women I feel so desperately sorry for the women because they pay rent in nearby townships, they pay up to Rand 400 a month rent, and they earn a pittance... So, they have, they often access contraception at the local clinics they are frightened to go, they don't report rape. So, they're totally at the mercy of what of whatever men actually use their services for the crushing but not all the men use the woman to do that crushing others use what they call the *pendukas*. So, yes the women are the ones I feel the most sorry for."—Carina, key informant

During a conversation with a key informant in the artisanal mining sector, the following was mentioned:

> "There are women who are working there against their will. There are women that are being sexually violated in the space of Zama Zamas."—Tessa, key informant

In addition, women who are part of the artisanal mining sector in South Africa face abuse from the police, who often coerce them into sexual favors to avoid arrest. Many of these women are unaware of their rights and lack knowledge about the laws protecting them from gender-based violations (Action Aid, 2019). On the matter of the police, a group of women Zama Zama stated the following about challenges they face:

> "Only the police but for now they have stopped. But amongst us there is no problem."—Group of women Zama Zama

Another woman, Zama Zama, confirmed that the only challenges they face are with the police.

> "The only problem that we have is, if the police comes here. They can come, like the other day, I was standing here. The one was started to come this side. The owners of the pundakas, they pay the police for them to go, so that they can not take our things. After that they come again, another one. They need again money. And then if we complain, they take our stuff and go. They go and sell it to another place, they don't take it to the station. That is corruption."—Esha, Zama Zama

Often, women are arrested for no reason, their equipment and minerals are confiscated, and the police take advantage of the fact that women, in most cases, do not know their rights. Furthermore, in some instances, mediators take advantage of women in mineral sales since women do not have equal access to inputs for their operations (Action Aid, 2019).

A Way Intended for Economic Empowerment?

Despite the aforementioned challenges and risks, data collected during the field visits in the Gauteng Province, South Africa, also shows that women Zama Zamas are using artisanal mining to create a more robust economic base for themselves. The money they earn through the Zama Zama sector is enough to sustain their families. One Zama Zama woman explained that she left her domestic job in Krugersdorp and moved to Soweto, where she is currently a Zama Zama miner:

> "I have been working in Krugersdorp, but the money it was sick. It was too small. By the time I was working there I couldn't afford to pay kids for school fees, and rent and transport, and even food. I was suffering. That is why I decided to leave. And then I started this. But since I have started to work this, everything is going well, back to normal. Very well. Yesterday, I sell my little gold, it was two gram point four. So when I sell it, I sell it for R850. I made R1980. Today, my boys are still working, they are busy. I will collect, I am going to sell it again, today. Yesterday I made R1980 today I will make 2 something. That is why I am saying I can't afford to work for the whole month for R1500. I have three kids, I have my mother, my farther they are still alive. I have to feed them, and my inlaws they are still alive."—Esha, Zama Zama

Another woman Zama Zama who, at that time, had only recently started, said the following:

> "It is my first time to this. I don't know if it will sustain me and the kids. I am trying to make a living because I see other women making a living out of this. I am still learning, and I am still new, but I saw no other options because I have been trying to find work and did not get work. So I ended up deciding to do this."—Linda, Zama Zama

In contrast, in terms of earnings, the literature reports that there is often a significant difference between what female artisanal miners earn compared to their male counterparts. For example, women working as artisanal miners in the DRC often work 12 h daily for a salary of 1000 Congolese francs (USD 1) daily (Peyer & Maillard, 2011). It has been reported that women earn less money than men despite working more hours. This often leads to women seeking other sources of income, amongst other things, prostitution (Environmental Law Institute, 2014). Working with dangerous substances such as mercury and arsenic is common in processing. Despite these risks, women are paid less for this work than men. Even if men are involved in processing work, they will most likely be paid more than their female counterparts (Ofosu et al., 2022).

However, during fieldwork discussions with women, it was mentioned that they earned between ZAR 700 and ZAR 900 per gram (approximately USD 37 to USD 48 as of September 28, 2024). It was confirmed that the gold buyer determines the price range and this is not influenced by gender.

Surprisingly, in some interviews with women in the Zama Zama mining sector, they mentioned facing fewer challenges than what was expected based on earlier accounts and the literature. When asked about challenges, the interviewees were unanimous in their experiences that they face few challenges. The women responded as follows:

> "There are no challenges. I just recently started. We are making a living and feeding our kids out of this. I haven't come across any issues because I do not go underground. I only work on the surface"—Linda, Zama Zama

> "I don't have problem about this. I don't want to lie. Because everything is going smooth".—Esha, Zama Zama

A possible explanation for these varied responses is the complexity of the Zama Zama sector. During a group discussion with key informants, it was explained that Zama Zama sites differ significantly, and the challenges women face depend on factors such as demographics, ethnicity, and locality. One key informant explained it to me as follows:

> "My thoughts are different areas work differently since we visit different areas all the time where the Zama Zama miners are working. So yesterday we were at Matholeville. Matholeville is a very calm space. It's calm with the Zama Zamas as well. The only destruction that Matholeville. Normally, it's when the Basuto nations from other areas come and want to fight with them for their gold. That is when problems start. As we heard yesterday, they were explaining that even last week, there was a shooting where two people were shot and they went to hospital. Fortunately enough, no one died, but there was a shooting, and that was resorted by the security Zama Zamas calling the police officials that they work with."—Ross, key informant

The insight gained from these empirical findings indicates that women cannot all be treated the same nor the same way as men. This finding has important implications for developing initiatives around the sector. As the findings indicate, the set of challenges women face is different and is informed by various factors, such as the artisanal mining site and the cultural and ethnic makeup of these sites, to mention a few.

To conclude, some studies have argued that the mining reforms have negative consequences for women and that the presentation of conflict minerals and the GBV narratives have provided an inadequate framework for developing practical solutions to the problems and challenges women face in ASM (Autesserre, 2012; Kelly et al., 2014). Although the problem of GBV and sexual assault is a severe one, the one-dimensional emphasis on GBV and sexual assault has risked obscuring the complexity of gender dynamics in ASM, particularly in eastern DRC (Bashwira et al., 2014; Bashwira & van der Haar, 2020; Laudati & Mertens, 2019; Maclin et al., 2017). Also, it is concerning that policymakers often fail to recognize the economic significance of ASM for women's livelihoods, even in conflict zones (Bashwira et al., 2014).

In a similar vein, the same can be said of the artisanal mining sector in South Africa, where there is an exclusive focus on hardship and the set of challenges the women face. Vital to note is that it is a reality for many women, as seen from the literature and research participants' quotes. However, there is also a contrasting narrative where the sector provides women with economic opportunities, enabling them to earn more than they would through domestic work, as expressed by themselves.

Conclusion

A narrow focus on women's experiences limits women to victims, disregarding the active role women voluntarily seek and play in mining (Maclin et al., 2017). The singular focus on sexual violence masks the economic significance of ASM as a livelihood activity for women (Laudati & Mertens, 2019). This reasoning is supported by the research of Bashwira et al. (2014), showing that a narrow focus on sexual violence in the sector risks muddying the nuanced dynamics of gender in artisanal mining. At least until recently, officials did not consider the prospect that small-scale and artisanal mining can serve as *"an economic stepping stone for women,"* even in conflict areas like the eastern Democratic Republic of the Congo (Bashwira et al., 2014). Women's motivations for living in and around mines are not taken seriously, and their direct and indirect contributions to the artisanal mining sector are routinely disregarded (Bashwira et al., 2014).

The literature contains numerous accounts of sexual violence, physical abuse, and other forms of discrimination against these women (see International Trade Union Conference (ITUC), 2011; Hilson et al., 2018). On the one hand, throughout sub-Saharan Africa, women are taking enormous risks to engage in ASM and, once involved, expose themselves to even more dangers. Conversely, despite facing significant challenges, numerous women engaged in ASM perceive the risks as a worthy endeavor and even regard it as a way to economic empowerment.

This perspective is not exclusive to women in the artisanal mining sector, as will be explored in the following chapter. In many cases, the socioeconomic opportunities provided by artisanal mining surpass the significant challenges that artisanal miners encounter. The next chapter unpacks the diverse challenges and opportunities in the artisanal mining sector, by highlighting the sector's impacts.

References

Action Aid. (2019). *Regularising informal artisanal mining in South Africa: An evidence-based report*. Action Aid.

Arthur-Holmes, F., & Busia, K. A. (2022). Safety concerns and occupational health hazards of women in artisanal and small-scale mining in Ghana. *The Extractive Industries and Society, 10*, 101079. https://doi.org/10.1016/j.exis.2022.101079

Autesserre, S. (2012). Dangerous tales: Dominant narratives on the Congo and their unintended consequences. *African Affairs, 111*(443), 202–222. https://doi.org/10.1093/afraf/adr080

Bafilemba, F., & Lezhnev, S. (2015). *Congo's conflict gold rush. Bringing gold into the legal trade in the Democratic Republic of the Congo*. Accessed March 1, 2016, from http://www.enoughproject.org/files/April%2029%202015%20Congo%20Conflict%20Gold%20Rush%20reduced.pdf

Bashwira, M.-R., Cuvelier, J., Hilhorst, D., & Van Der Haar, G. (2014). Not only a man's world: Women's involvement in artisanal mining in eastern DRC. *Resources Policy, 40*, 109–116. https://doi.org/10.1016/j.resourpol.2013.11.002

Bashwira, M. R., & van der Haar, G. (2020). Necessity or choice: Women's migration to artisanal mining regions in eastern DRC. *Canadian Journal of African Studies / Revue Canadienne Des Études Africaines, 54*(1), 79–99. https://doi.org/10.1080/00083968.2019.1674170

Beneria, L. (2010). Globalization and gender: Women's labor in the global economy. In B. Berberoglu (Ed.), *Globalization in the 21st century*. Palgrave Macmillan. https://doi.org/10.1057/9780230106390_8

Bester, V. (2019). *A corporate social responsibility (CSR) conceptual framework to address unregulated artisanal gold mining in South Africa*. Doctoral Thesis / Master's Dissertation. University of Johannesburg. Accessed November 3, 2024, from https://ujcontent.uj.ac.za/esploro/outputs/doctoral/A-corporate-social-responsibility-CSR-conceptual/9911598107691

Burke, G. (1997). Policies for small-scale mining: The need for integration. *Minerals & Energy - Raw Materials Report, 12*(3), 11–14. https://doi.org/10.1080/14041049709409441

Buss, D., Rutherford, B., Kumah, C., & Spear, M. H. (2021). Beyond the rituals of inclusion: The environment for women and resource governance in Africa's artisanal and small-scale mining sector. *Environmental Science & Policy, 116*, 30–37. https://doi.org/10.1016/j.envsci.2020.10.019

Cuvelier, J. (2014). Work and masculinity in Katanga's artisanal mines. *Africa Spectrum, 49*(2), 3–26. https://doi.org/10.1177/000203971404900201

Danielsen, K., & Hinton, J. (2020). A social relations of gender analysis of artisanal and small-scale mining in Africa's Great Lakes Region. *Canadian Journal of African Studies, 54*(1), 17–36. https://doi.org/10.1080/00083968.2019.1676807

Davidson, J. L. (1993). The transformation and successful development of small-scale mining enterprises in developing countries. *Natural Resources Forum, 17*(4), 315–326. https://doi.org/10.1111/j.1477-8947.1993.tb00192.x

Environmental Law Institute, The. (2014). *Artisanal and small-scale gold mining in Nigeria. Recommendations to address mercury and lead exposure*. Accessed March 1, 2016, from http://www.eli.org/sites/default/files/eli-pubs/nigeria-asgm-assessment-final-report.pdf

Hilson, G., Hilson, A., Siwale, A., & Maconachie, R. (2018). Female faces in informal 'Spaces': Women and artisanal and small-scale mining in sub-Saharan Africa. *Africa Journal of Management, 4*(3), 306–346. https://doi.org/10.1080/23322373.2018.1516940

Hinton, J. J., Veiga, M. M., & Beinhoff, C. (2003). Women and artisanal mining: Gender roles and the road ahead. In G. Hilson (Ed.), The Socio-Economic Impacts of Artisanal and Small-Scale Mining in Developing Countries (pp. 161–203). A.A. Balkema, Swets Publishers.

Hinton, J. (2016). *Advancing gender equality in the scaling up minerals traceability project.* Accessed January 2, 2023, from https://www.itsci.org/wp-content/uploads/2018/02/GRF-Final-Report-Pact-ASM-and-Gender-Scaling-Up-Minerals-Trace.pdf

Hinton, J., Hinton, B. E., & Veiga, M. M. (2006). Women in artisanal and small scale mining in Africa. In K. Lahiri-Dutt & M. Macintyre (Eds.), *Women miners in developing countries: Pit women and others* (1st ed.). Routledge. https://doi.org/10.4324/9781315233734-11

Hinton, J., & Mbabazi, D. (2009). Women and artisanal mining: Gender roles and the road ahead. In G. Hilson (Ed.), *The socio-economic impacts of artisanal and small-scale mining in developing countries* (pp. 161–203). CRC Press.

International Trade Union Conference (ITUC). (2011). *Violence against women in Eastern Democratic*. Accessed March 20, 2020, from https://www.ituc-csi.org/IMG/pdf/ituc_violence_rdc_eng_lr.pdf.pdf

Intergovernmental Forum on Mining, Minerals, Metals and Sustainable Development (IGF). (2018). Women in Artisanal and Small-Scale Mining: Challenges and opportunities for greater participation. Winnipeg: IISD.

Jenkins, K. (2014). Women, mining and development: An emerging research agenda. *The Extractive Industries and Society, 1*(2), 329–339. https://doi.org/10.1016/j.exis.2014.08.004

Kelly, J. T. D., King-Close, A., & Perks, R. (2014). Resources and resourcefulness: Roles, opportunities and risks for women working at artisanal mines in South Kivu, Democratic Republic of the Congo. *Futures, 62*, 95–105. https://doi.org/10.1016/j.futures.2014.04.003

Labonne, B. (1996). Artisanal mining: An economic stepping stone for women. *Natural Resources Forum, 20*(2), 117–122. https://doi.org/10.1111/j.1477-8947.1996.tb00644.x

Lahiri-Dutt, K. (2008, November 29). *Digging to survive: Women's livelihoods in South Asia's small mines and quarries*. https://ssrn.com/abstract=1716580.

Lahiri-Dutt, K. (2015). The feminisation of mining. *Geography Compass, 9*(9), 523–541. https://doi.org/10.1111/gec3.12229

Laudati, A., & Mertens, C. (2019). Resources and rape: Congo's (toxic) discursive complex. *African Studies Review, 62*(4), 57–82. https://doi.org/10.1017/asr.2018.126

Maclin, B. J., Kelly, J. T. D., Perks, R., Vinck, P., & Pham, P. (2017). Moving to the mines: Motivations of men and women for migration to artisanal and small-scale mining sites in Eastern Democratic Republic of the Congo. *Resources Policy, 51*, 115–122. https://doi.org/10.1016/j.resourpol.2016.12.003

Malpeli, K. C., & Chirico, P. G. (2013). The influence of geomorphology on the role of women at artisanal and small-scale mine sites. *Natural Resources Forum, 37*(1), 43–54. https://doi.org/10.1111/1477-8947.12009

Meinzen-Dick, R., Johnson, N., Quisumbing, A. R., Njuki, J., Behrman, J. A., Rubin, D., Peterman, A., & Waithanji, E. (2014). The gender asset gap and its implications for agricultural and rural development. In A. R. Quisumbing et al. (Eds.), *Gender in agriculture: Closing the knowledge gap* (pp. 91–115).

Ofosu, G., Torbor, M., & Sarpong, D. (2022). Gender and artisanal and small-scale mining: Exploring women's livelihood and occupational roles in formalised settings. *Journal of Rural Studies, 96*, 121–128. https://doi.org/10.1016/j.jrurstud.2022.10.013

Olayide, P. E., Olawoye, J. E., & Olayide, O. E. (2013). Gender dimensions of rural livelihoods in artisanal and smallscale mining in Itesiwaju local government area of Ogo State, Nigeria. *African Journal of Sustainable Development, 3*(1), 72–87.

Perks, R., Kelly, J. T. D., Constantian, S., & Pham, P. (2018). Resources and resourcefulness: Gender, human rights and resilience in selected artisanal mining towns of Eastern Congo: Informal, artisanal and small-scale mining in the contemporary world. In K. Lahiri-Dutt (Ed.), *Between the plough and the pick: Informal, artisanal and small-scale mining in the contemporary world* (pp. 209–231). ANU Press. https://doi.org/10.22459/BPP.03.2018.10

Peyer, C., & Maillard, Y. (2011). *Contracts, human rights and taxation: How a company exploits a country. The case of glencore in the Democratic Republic of Congo*. Accessed February 3, 2016, from http://www.bench-marks.org.za/research/drc_study_english_edition_6.pdf

Traore, P. A. (1994). Constraints on small-scale mining in Africa. *Natural Resources Forum, 18*(3), 207–212. https://doi.org/10.1111/j.1477-8947.1994.tb00889.x

United Nations (UN). (1996). Recent developments in small-scale mining. *Natural Resources Forum, 20*(3), 215–225. https://doi.org/10.1111/j.1477-8947.1996.tb00655.x

United Nations Security Council. (2015). *Final report of the group of experts on the Democratic Republic of the Congo*. Accessed February 8, 2016, from http://www.securitycouncilreport.org/atf/cf/%7B65BFCF9B-6D27-4E9C-8CD3-CF6E4FF96FF9%7D/s_2015_19.pdf

World Health Organisation (WHO). (2016). *Artisanal and small-scale gold mining and health*. Accessed January 14, 2019, from 9789241510271-eng.pdf (who.int)

World Bank. (2024). *State of the Artisanal and Small-Scale Mining Sector 2023*. World Bank. Accessed September 28, 2024, from https://www.worldbank.org/en/news/press-release/2024/02/05/world-bank-urges-action-for-gender-equality-in-artisanal-and-small-scale-mining

Yousafzai, M. (2013). *Malala Yousafzai: 16th birthday speech at the United Nations*. Accessed November 15, 2024, from https://malala.org/newsroom/malala-un-speech

Chapter 6
The Bright Side, the Dark Side, and the Ramifications of Mining

For mines are for men, not for money. And money is not something to go mad about, and throw your hat into the air for. Money is for food and clothes and comfort, and a visit to the pictures. Money is to make happy the lives of children. Money is for security, and for dreams, and for hopes, and for purposes. Money is for buying the fruits of the earth, of the land where you were born.

—Alan Paton (1948)

Introduction

Artisanal mining, often hailed as a path to economic empowerment and local development, unfolds against a backdrop of undeniable complexities. On one hand, artisanal mining represents a lifeline for numerous individuals and communities, offering opportunities for self-reliance and income generation. The artisanal and small-scale mining (ASM) sector can be regarded as a viable pathway for creating sustainable livelihoods and may serve as a poverty alleviation initiative, specifically in sub-Saharan Africa (SSA) (Osumanu, 2020; Mhangara et al., 2020).

In South Africa, artisanal mining is often the only income source for an individual or a household. The Zama Zama route became a way to make a livelihood since limited job opportunities exist in the formal sector (Action Aid, 2019).

Conversely, the artisanal mining sector is not without its challenges. It often operates in the shadows of formal regulation, leading to issues of environmental degradation, unsafe working conditions, and, sometimes, exploitative practices. These challenges starkly contrast the industry's potential for positive socioeconomic impact. Unfortunately, governments and industry players often neglect this sector by focusing on the negative impacts of ASM rather than addressing the challenges to improve the sector's sustainable development opportunities (Buxton, 2013).

This chapter unpacks the complexities of artisanal mining, delving into the social and environmental impacts it frequently causes. The first part of this chapter will offer an overview of the social and environmental issues often intertwined with the artisanal mining sector. It is crucial to recognize both the potential benefits and the significant challenges that are part of artisanal mining. Ignoring these challenges could hinder its development and limit the socioeconomic benefits for the artisanal

V. Bester, *The Untold Story of Zama Zama Miners in South Africa*, Clinical Sociology: Research and Practice, https://doi.org/10.1007/978-3-031-82783-9_6

miners involved. The latter part of the chapter will juxtapose the social and environmental consequences of artisanal mining with those associated with large-scale mining (LSM) operations.

Health and Working Conditions of the Artisanal Mining Sector

Artisanal mining is often associated with several negative impacts, for example, issues related to the miners' health and safety, the environment, and violence and conflict within the sector. Furthermore, ASM communities can be urban but are usually rural and already operating on the fringes of society, making access to healthcare facilities tricky and nearly impossible. Poor living conditions often characterize artisanal miners' communities. Furthermore, hygiene and sanitation conditions are poor, access to clean water is limited, and diseases like tuberculosis and Human Immunodeficiency Virus/Acquired Immunodeficiency Syndrome (HIV/AIDS) are common. Buxton (2013) explains that artisanal miners face hazardous working conditions as they: *"work in unsupported tunnels (of 1.5 meters in diameter and up to 90 meters deep) drilling and removing rock with hand tools and carrying the ore to the surface in sacks."*

In South Africa, the miners work under circumstances similar to Buxton's description. Most of the time, artisanal miners utilize no protective gear, increasing the danger level. Also, specific health and safety threats depend on the type of ASM that occurs.

For example, the gold artisanal mining sector is notorious for its severe dangers and negative health impacts. Action Aid (2019) documented that when the Zama Zama miners were asked about their working conditions, most of the respondents working in gold mining sites were unhappy with the working conditions, in contrast to the diamond mining sector, where most of the respondents (90%) were satisfied with their working conditions (Action Aid, 2019).

Apart from the difference in ore, the sites where artisanal mining occurs also differ, posing diverse health and safety risks. In South Africa, artisanal mining occurs on the surface and underground in mines that are decommissioned, abandoned, or still operational (Minerals Council South Africa, 2023). Artisanal miners have been reported to stay underground for days or weeks, facing inhumane circumstances, hoping to find minerals to make a living (Hosken, 2017).

Also, almost all artisanal miners face ergonomic stresses, joint and muscle pains, hearing loss, and exposure to dust containing hazardous chemicals. Many face the risk of accidents resulting in severe injury or even death (De Haan et al., 2020). Between 2012 and 2015, at least 312 Zama Zama miners died from mining-related accidents in South Africa (Snell, 2019). This mortality rate is believed to be underreported by the Zama Zamas, who rarely trust local authorities (Snell, 2019). Besides the high risks the miners face, they do not have access to Personal Protective Equipment (PPE), nor do they have the means to access funding, skills development, and proper technology to make the circumstances safer (Action Aid, 2019).

During interviews with Zama Zamas, they expressed concerns about their health and safety. A Zama Zama explained the situation as follows:

> "The biggest risk by now is death. People are dying. Why people are dying, people they don't have knowledge, they don't know actually where it's dangerous. Where there is barricade, where there is no entry no go area. We are using these places that's closed for a long time you see."—Tim, Zama Zama

Martin, another Zama Zama, agreed with his colleague:

> "The risks. Underground you go there is big rocks. So, some rocks there are cracks so us we are just checking for gold; if you touch wrong the rock is going to what? It's going to fall down. So that you are going to get dead or a big injury. You understand?"

When probing a key informant during a conversation on some challenges, the following was mentioned:

> "My goodness, the major challenges. I don't think I can put them in any, in any particular [order], but it is the danger underground. So, its theft. It is their personal safety, personal safety whether it be underground or above ground. That's the biggest, the biggest challenge."—Carina, key informant

Despite these dangers, Zama Zama miners continue to mine under these conditions. For them, artisanal mining is a means to create a livelihood.

In terms of miners' health and safety, these highly unsafe working conditions are not the only issue; mercury usage in artisanal mining is widespread.

In artisanal mining, mercury extracts the gold from the ore by forming an amalgam, a 50:50 mixture of mercury and gold. The amalgam is heated, and the toxic mercury evaporates, leaving the gold behind. In 2019, mercury was sold at ZAR 350 per bottle, and the miners can process 50 gram of gold from one bottle of mercury (Action Aid, 2019). It has been documented that mercury is procured through syndicate connections (Martin, 2019).

The type of mercury exposure that the miners face is regarded as severe and acute because the dosage that gets inhaled by the artisanal miners is very high and mainly causes damage to the kidneys and lungs, which may potentially damage the nervous system (Richard et al., 2014). Despite the negative impacts of mercury usage in artisanal and small-scale gold mining, it is still widely used. Because it is inexpensive, effective in the gold process, easy to use, and easily accessible by miners, who can use it independently (Persaud & Telmer, 2015).

Mercury exposure affects artisanal miners and the surrounding communities living and working close to the artisanal mining sites (UNEP, 2012). The Environmental Law Institute (2014) described this pathway as particularly harmful because more mercury enters the body through breathing in the vapors than through ingestion or direct contact. A civil society worker stated the following on the use of mercury by the Zama Zamas:

> "The third area is because they do an elementary processing of the material that they bring up, and they don't have access to highly mechanized procedures, and so on, they use the, sort of very early days of … mercury to extract the metals. And that's at huge, huge risk to their health."—Jane, civil society

Artisanal miners are often unaware of the negative impacts of mercury usage (Chupezi et al., 2009). The following extracts portray mercury usage during artisanal mining and the miner's ignorance of the dangers.

> "Ja, there is a daga that are breaking that stones nê. After that, you crush it to make it fine, and you put water to make daga nê, after that you put mercury inside and start to squeeze it so that mercury is going to do what, gold."—Martin, Zama Zama

Also, consider this exchange:

> Researcher: "Do you use mercury in your processing?"
> Zama Zama: "Yes."
> Researcher: "Do you wear anything to protect yourself when you use the mercury?"
> Zama Zama: "No"
> Researcher: "And a mask?"
> Zama Zama: "Ja, no mask."—Sam, Zama Zama

The health and safety issues within artisanal mining serve as a driver to portray the sector in an exclusively negative light. Also, the sector may include other ills such as child labor, violence, conflict, and environmental damage. The following sections examine these impacts often associated with artisanal mining.

Child Miners

According to the International Labour Organisation (ILO) (2011), there are an estimated 215 million child laborers worldwide, of which more than half (53%) do hazardous work, like artisanal mining. Sub-Saharan Africa has the highest incidence of child labor (one in five children), with an estimated 59 million children engaged in child labor and nearly 29 million in hazardous work (O'Driscoll, 2017).

It is essential to note that it is difficult to accurately determine the number of children involved in the ASM sector due to the overall data gap (as discussed in Chap. 4). Moreover, the nature of child labor is often a secret because it is illegal, making estimations even more challenging.

The ILO (2011) confirmed that most child miners work in the ASM sector. The ILO estimated in 2006 (nearly 20 years ago) that the number of child miners was around 1 million. The exact figure was repeated in a 2019 report from the ILO (ILO, 2019). What is concerning is that the WHO (2016) conveyed that it is nearly impossible to eradicate children's participation in ASM because of this sector's strong family ties, its temporary and informal nature, and high poverty levels.

A 2021 article by the South African-based Institute for Security Studies reported that children mine a considerable amount of the Democratic Republic of the Congo's (DRC's) coltan. The report estimates that over 40,000 child and teenage miners are mining coltan in the DRC (Ojewale, 2021). Coming from remote villages and towns in Kivu, children either drop out of school or have never had the opportunity to attend.

Coltan is used in cell phones, laptops, and other devices because of its particular ability to store and release electrical energy. In 2019, 40% of the global coltan supply was produced in the DRC. As 5G technology grows, the demand for coltan will increase.

It is perhaps worth mentioning that the DRC is a signatory of notable international certification protocols in its pursuit of combating child labor. These protocols include the Regional Certification Mechanism of the International Conference of the Great Lakes Region, the Kimberley Process Certification Scheme, and the Dodd-Frank Act. The country is also a signatory to the Organization for Economic Co-operation and Development Due Diligence Guidance for the responsible processing of minerals from areas in conflict and the Conflict Minerals Regulation. However, still, children are being exploited in the DRC's coltan mines. The issue is that mineral certification schemes are neither adequately implemented nor the law regarding child labor and associated hazardous working conditions. While the frameworks are complementary, their feasibility in the DRC is questionable (Ojewale, 2021).

Furthermore, child labor is a serious risk, especially regarding downstream companies' reputations (Deberdt, 2021). Research in the DRC has shown that every on-the-ground participant, from civil society and service providers to Congolese government officials and representatives of foreign organizations, agreed that child labor has dominated discussions (Deberdt, 2021). While acknowledging the existence of child labor, they contend that it is a symptom of more significant problems that have not been addressed, such as widespread poverty, a lack of viable economic options, and the absence of educational opportunities (Deberdt, 2021).

Children who work in mine pits and excavations are regarded as the poorest of the poor. The ILO (2011) described poverty as one of the key drivers of child labor.

The ILO (2011) has documented a list of common mining and quarrying tasks for children in artisanal mining alongside related hazards and consequences. Child miners fulfill various tasks, including ore transportation, rock crushing, and the processing of gold by mixing it with mercury (Human Rights Watch, 2013). Child miners can be intensively involved in mining by fulfilling tasks such as digging or handpicking ore to amalgamate the ore. The children could develop various health problems, such as respiratory problems, silicosis, neurological damage, and musculoskeletal disorders, to name a few. Since artisanal mining involves mercury, children are exposed to high mercury poisoning through contact with mercury and inhaling mercury fumes (Human Rights Watch, 2013). The dangers are that mercury may cause developmental and neurological problems in children because it attacks the central nervous system (Human Rights Watch, 2013).

Because of these unsafe circumstances, child miners may be involved in other tasks, including cooking for adult miners, transporting minerals or ore, and selling goods and services to miners. These tasks put child miners at risk for physical and sexual abuse. Lastly, children participating in artisanal mining risk missing educational opportunities and experiences (Human Rights Watch, 2013). Table 6.1 provides an overview of the key dangers and risks faced by children who work in artisanal mining.

Table 6.1 Dangers and risks faced by children in artisanal mining and their consequences (ILO, 2019)

Task	Dangers and risks faced by child miners	Consequences
Digging; hand-picking ore, slaps, or rocks; tunnelling; and going down into tunnels	The use of heavy tools and explosives; carrying of heavy tools, heavy loads, doing repetitive movements; exposure to dangerous heights and open holes; exposure to confined spaces with faulty support structures, exposure to radiation, poisonous gasses, dust, and damp	Children die or get injured from tunnel collapse; they suffocate from compressor mining; get injured from explosions; develop silicosis and related respiratory diseases; children get exhausted from this labor-intensive work. The child miners develop joint and bone deformities; blisters on hands and feet; cuts and slashes on the body; back and muscle injuries occur; head trauma; noise-induced hearing loss; respiratory difficulties are common; frostbite, sunstroke, and dehydration
Crushing, washing, and amalgamating Removing waste or water from mines and transporting of minerals	The usage and exposure to lead, mercury, and other heavy metals; extreme exposure to dust and carrying of heavy loads	The child miners have a great chance to develop neurological damage; genitourinary disorders; musculoskeletal disorders, fatigue, immune deficiency
Cooking and cleaning for adults and selling goods and services to miners	Exposure to unsafe stoves, explosive fuels, physical and verbal abuse	The child miners can experience injury from beatings, sexual abuse. This can lead to the development of behavioral disorders
Mining and quarrying in general	Remote locations; lawless atmosphere; poor sanitation, contaminated drinking water, stagnant water and mosquitoes, inadequate nutrition, recruitment into the sex trade, gambling, drugs, and alcohol	Child miners can die because of a lack of medical treatment. They can also develop behavioral disorders; addictions; sexually transmitted diseases; unwanted pregnancies; stunted growth; diarrhea and digestive disorders; malaria and mosquito-borne diseases

In the South African context, it was highlighted that the involvement of children in artisanal mining is an issue. Also, children are often indirectly involved, putting them at risk. In support of this, Carina, a key informant, said the following about women and their children during an interview:

> "…and very often a lot of them have got their children with them. Babies on their backs with that dust…"

A thorough baseline study should be conducted to determine the prevalence of child miners on any site. It is further necessary to examine any influencing factors, such as the availability of education resources and poverty levels in a community, to establish the driving characteristics of child miners in any given context.

Violence and Conflict Linked to Artisanal Mining

Artisanal mining is often associated with violence and conflict. The WHO (2016) indicated that where artisanal and small-scale gold mining operations are regarded as illicit, inter-gang war can cause an upsurge in violence between miners, authorities, and local land users. One interviewee also described this during a discussion.

> "So, what is happening is, in the communities, the reports are such that there are a lot of violence that comes from not the artisanal miners themselves but the criminal gangs that wants to take advantage of the proceeds of the artisanal miners. So what they do, they would go to the houses of the known artisanal miners. Of the Zama Zamas, and demand whatever they got."—Mike, civil society

Violence and conflict are common in the Zama Zama sector. For example on March 21, 2023, the media reported that the livelihoods of 200 waste pickers hang in the balance as a bloody turf war between Zama Zamas, which has reportedly claimed several lives, has left many in fear (Singh & Makhetha, 2023). Early in March 2023, the Simmer and Jack Landfill in Germiston closed for a week after Zama Zamas threatened the reclaimers (waste pickers) if they did not each pay a *"protection fee."* The landfill was constructed on the Simmer and Jack Gold Mine, closed in 1969. The article states that a resident of the adjacent Summerfield community reports hearing gunfire echo through the landfill every evening. The respondent, as per the article, was quoted saying: "*We hear gunshots every night, and it's not just one gunshot; it is multiple at a time. Sometimes you see men carrying AK47's standing on top of the hill at the landfill site. It's scary because we live just across the road from where this is happening, and I have small children.*" As per the article, a private security guard said it was too risky to enter the landfill, even for police.

Furthermore, an element that fuels violence in the sector is syndicates. During discussions, two civil society workers raised the matter and said the following:

> "Thirdly, they face issues of gangsterism. People now know when they go down there and they come back, they have gold. And they are being attack by these gangsters and taken all their gold."—Mike, civil society

> "Syndicates and gangs have moved in and are controlling the buying of gold and other minerals that are bought by small-scale and survival miners, that are mined by small-scale and survival miners. And this then has led to territory wars and people being shot and killed."—Danny, civil society

Another element that adds to the violence in this sector is the sporadic clashes with the formal mining industry. It is argued that the overall ASM sector suffers from critical social issues and ongoing conflict with large-scale mines, mainly because of its questionable legal status (Andrew & Hilson, 2003). Martin (2019) rightly remarks that, although active mines are legally entitled to remove Zama Zamas from the site, the current approach of using security and brute force to end the activity is misguided and counterproductive. Additionally, Stewart et al. (2019) describe Zama Zama mining's negative impact on the formal mining sector by examining its effects on the formal miner's health and safety.

Since policing is seen to counteract artisanal mining, police raids contribute to conflict at mine dumps (Jinnah, 2016). Also, miners are often mercilessly exploited by police, government representatives, and criminal gangs (Thornton, 2014). When miners are arrested, they pay a fine for trespassing, and the charges are often dropped (Jinnah, 2016). More commonly, miners bribe the police to be released after being arrested (Jinnah, 2016). Despite these challenging working conditions, the Zama Zama miners continue their work in the hope of earning money.

The Environmental Impact of Artisanal Mining

Of all the artisanal mining issues, the environmental problem (specifically mercury usage) has received much attention. Artisanal and small-scale gold mining (ASGM) is regarded as the primary source of mercury in the global atmosphere, accounting for 37% (727 tons) of all global emissions (Rajee et al., 2015). Mercury is an element that does not degrade in the environment. ASGM activities occur worldwide, causing global mercury releases and affecting countries in the Global South the most.

Fortunately, several programs and platforms have been developed to educate and facilitate artisanal miners in no mercury usage (United States Environmental Protection Agency, 2024; United Nations Environmental Programme, 2012). One such program, called *PlanetGOLD*, is led by the United Nations Environmental Programme in conjunction with the Artisanal Gold Council. The program aims to address mercury pollution in this sector by offering a platform where the public and private sectors can work together to support ASGM communities. This partnership has developed a practical guide on reducing mercury in ASGM with alternative methods. The step-by-step guide states that it can be used by policymakers, miners, and civil society. Adding to this is that mining companies can also use the guide and offer training to artisanal miners. Thus, building upon the capacity of these miners will inevitably aid in creating a more sustainable artisanal mining sector. This practical guide has been designed for private sector usage and provides the baseline to measure progress in transitioning away from mercury use (Persaud & Telmer, 2015).

Besides the negative impact of mercury on the environment, other environmental impacts also occur, such as erosion and land disturbance (Rajee et al., 2015). By removing the vegetation and topsoil of land also reduces an area's agricultural and ecological productivity (Rajee et al., 2015).

Land disturbance caused by ASM is another negative environmental impact (Hagos et al., 2016). Artisanal gold mining operations often involve the digging up of river channels and banks and surface trenching.

The lack of environmental regulation and compliance around ASM exacerbates water pollution, making artisanal miners contributors to water contamination. Because ASGM uses water for panning and processing, water canals are affected (Mujere & Isidro, 2016). These tailings often contain mercury and heavy metals, such as lead, cadmium, arsenic, and inorganic compounds. It has been widely documented that these metals seriously threaten human health and negatively impact food quality, groundwater, micro-organisms activity, and plant growth. In this regard, Rajee et al. (2015) stated that the water acidity in gold mining is of a low pH, thus affecting marine life and people's health via contamination, siltation, and extreme water withdrawals. However, miners often lack the knowledge, resources, or prerequisites to work in an environmentally friendly manner.

Artisanal mining is not the only sector that causes social and environmental issues. The following section looks at the social and environmental impacts of LSM operations; however, these concerns typically garner less media coverage compared to Zama Zama mining.

Large-Scale Mining and Its Impacts

Health and Safety

Mine and worker Occupational Health and Safety (OHS) remains crucial given the historically dismal record of fatalities and accidents in mining in South Africa (Stewart et al., 2019).

A total of 615 deaths were reported in 1994. This number dropped significantly over the years. A decade later, the total number of fatalities in 2014 was reported at 84 (Department of Mineral Resources, 2015; Stewart et al., 2019). While mining employment also declined, deaths dropped proportionately more than employment levels. However, from a historic low of 73 fatalities in 2016, fatal accidents rose to 90 deaths in 2017. Before falling back to 81 in 2018, these deaths ended a sustained improvement of the decade before (Njini, 2018). Important to note is that these figures do not include the deaths of the Zama Zama miners (Stewart et al., 2019).

Since 2003, the mining industry has seen a consistent decline in fatalities. In 2003, there were 270 reported deaths, but by the end of 2023, the number of fatalities was reported at 55. The platinum and gold sectors remain the leading contributors to these fatalities, with 22 and 20 deaths reported in 2023, respectively. In contrast, the coal sector experienced only 7 fatalities in 2023, a significant decrease from 90 in 1993 and 31 in 2000 (African Mining, 2024).

There has been an unquestionable drop in fatalities in the former mining industry since 2003 (African Mining, 2024). However, a key informant explained that there is a disproportionate focus on artisanal miners' deaths while neglecting the broader health and safety problems affecting workers in formal mining operations.

The interviewee said that while the media and public discourse may highlight the lack of safety equipment and the resulting fatalities among artisanal miners, there is a significant, often unspoken issue with the health risks and fatalities occurring in LSM operations. These include deaths from smoke inhalation, explosions, and various unquantified health issues. The key informant's observation calls for a more comprehensive examination of safety in the mining industry.

> "You've heard a lot of them that have died of smoke and others that have died of explosions. You've had others who have died of various diseases that relate to health issues that are not quantified. But nobody has spoken about that. Everybody is saying, oh there are 5 artisanal miners who died because they went there, and they didn't have safety equipment. We're not talking of the hundreds and perhaps thousands that are dying within the mining industry itself run by large mines."—Jack, key informant

Environmental and Social Impacts

ASM is not the only sector that negatively affects the environment; mining has various environmental impacts regardless of size and scale. The United Nations documented several significant environmental impacts of large-scale or industrialized mining (United Nations Economic Commission for Africa, 2011).

These examples are as follows: the social and ecological problems caused by mining can include acid mine drainage (AMD) and mining waste (United Nations Economic Commission for Africa, 2011). Also, there are overall negative consequences on the ecology caused by mining, for example, in the DRC forests (United Nations Economic Commission for Africa, 2011). LSM also poses dangers to human health from uranium mining and the extensive pollution of air and land (United Nations Economic Commission for Africa, 2011). Core mining activities mainly caused this pollution, like digging, pumping, discarding wastewater, and smelting operations that release sulfur dioxide.

Still, the media often fails to report or educate the public about the devastating effects of LSM on the environment and society. During the South African Human Rights Commission's consultations with communities affected by LSM operations, the most crucial factors raised concerning environmental impact were increased dust levels, safety concerns over blasting activities, and the effect of these factors on food security, health, and overall well-being.

Furthermore, allegations of pollution, lack of information, and insufficient levels of participation in matters concerning environmental management and impact were also raised by community members (SAHRC, 2018).

The coal mining industry is an excellent example of how mining companies caused substantial harmful ecological damage in South Africa. The Witbank coalfield, which accounts for 40% of South Africa's coal production, is the region in the world that is most contaminated by nitrogen dioxide.

In 2010, coal mining had already seriously degraded the Upper Olifants River that runs through the Limpopo and Mpumalanga provinces (Munnik, 2010). The Loskop Dam, part of a nature reserve in the Olifants River, has experienced a die-off of fish, turtles, and crocodiles resulting from poor water quality. The towns of Witbank and Middelburg are periodically too salinized to meet drinking water standards (Munnik, 2010). In May 2016, the Centre for Environmental Rights (2016) published a report on coal mining's environmental impact in Mpumalanga entitled *Zero Hour: Poor Governance of Mining and the Violation of Environmental Rights in Mpumalanga.* The report concludes that the government's failure to ensure that mining companies comply with the law is causing unprecedented environmental degradation and chronic health problems in Mpumalanga, with dire consequences for communities and South Africa's future prosperity. This situation results from neglect, limited resources, and willful inaction by the Mineral Resources and Water and Sanitation Departments. The coal mines of South Africa are not the only culprits when we talk about pollution.

In 2016, the International Human Rights Clinic from Harvard Law School published a report entitled: *The Cost of Gold: Environmental, Health, and Human Rights Consequences of Gold Mining in South Africa's West and Central Rand.* The report documents the threats posed by water, air, and soil pollution from mining in the West and Central Rand. AMD has contaminated water bodies that residents use to irrigate crops, water livestock, wash clothes, and swim. Also, dust from mine waste dumps has covered communities. Moreover, the government has allowed homes to be built near and sometimes on those toxic and radioactive dumps.

During conversations, one interviewee alluded to the notion of mining companies avoiding their responsibilities when it comes to environmental safeguarding:

> "Look I think that I can give you lists that I have on my computer of mines that don't have water licenses or mines that are not compliant. In fact I recently got a list of mines not complying with the law in basically in a variety of ways."—Danny, civil society

Someone from a civil society organization referred to the same issues by stating:

> "… some mines will operate without water license. Some of them will operate without social labor plans that have been done in consultations with the communities, some mining companies will have very interesting EIAs, environmental impact assessment projects that were really not done on the ground."—Brandon, civil society

The issue of AMD is linked to the coal and gold mining industry. AMD poses a significant environmental and public health threat in South Africa, particularly in the Witwatersrand Basin around Johannesburg. This toxic legacy of over a century of gold mining occurs when water interacts with exposed rock surfaces in abandoned or operational mines, generating sulfuric acid and releasing heavy metals (Business and Human Rights Resource Centre, 2024) The contaminated water then seeps into groundwater or flows into surface water bodies, polluting rivers like the Vaal and Crocodile and affecting downstream communities (Business and Human Rights Resource Centre, 2024).

The scale of the problem is alarming, with an estimated 278 abandoned mines and 200 mine dumps in the Johannesburg area alone, contributing to AMD.

The Zama Zamas, who are re-mining these dumps, are highly exposed to heavy metals as they handle AMD, often without protective gear. Contamination can spread up to 20 km from mine sites, affecting soil quality, urban agriculture, and human health (Mafata, 2024).

In Soweto, home to 1.9 million people, research has found that AMD has contaminated streams, irrigation water sources, and soil adjacent to the Klip River. Exposure to the heavy metals in AMD can cause a range of health issues, from mild symptoms like dehydration and abdominal pain to severe consequences such as birth defects, brain damage, and cancer (Mafata, 2024).

The South African government has acknowledged the seriousness of the AMD crisis and has begun implementing action plans, including water treatment facilities, but the long-term management of this persistent environmental challenge remains a significant concern (Business and Human Rights Resource Centre, 2024).

Artisanal miners frequently work without adequate safety measures or environmental protections, worsening the environmental damage from previous formal mining activities. The absence of closure certificates for large-scale mines, especially those with significant environmental impacts like gold and coal mines, exacerbates the issue. The government's failure to enforce closure regulations, combined with mining companies' tendency to sell mines to less well-resourced entities, has created a cycle of neglect, hindering efforts to promote sustainable and responsible mining practices.

Conclusion

The chapter highlighted that ASM can cause and intensify social and environmental problems. However, it also emphasized that this sector is not the sole contributor to such adverse effects by examining the harmful impacts of LSM operations on society and the environment. Also, exploring the tendency of LSM operations to neglect environmental safeguards reveals a clear link between the actions of mining companies and the rise of Zama Zama mining, particularly when these companies abandon their responsibilities regarding mine closure.

Acknowledging the economic opportunities alongside the social and environmental impacts of the Zama Zama sector is essential. In addition, exploring the socioeconomic opportunities of the South African artisanal mining sector in depth will help to address the sector's challenges. Osumano (2020) correctly observes that while ASM problems are widely acknowledged and articulated in policymaking circles, the challenges that ASM rural communities face are vastly understated. Understanding ASM challenges and identifying institutional barriers perpetuating unsustainable livelihoods in rural mining communities is the first step in addressing ASM problems. Exploring the socioeconomic opportunities of the South African artisanal mining sector in depth will help to address the sector's challenges.

Likewise, as this sector grows, it is vital to recognize its activities from a benefits-oriented perspective and understand its adverse impacts (Action Aid, 2019). Additionally, artisanal mining is linked to generating income, creating employment, promoting youth and women's empowerment, and fostering small business growth (Action Aid, 2019).

Given the social and environmental challenges surrounding Zama Zama mining, it is clear why the South African mining sector has consistently struggled to tackle this issue, frequently passing the responsibility to the government. While this deflection is somewhat justified, there have been increasing calls in recent years to formalize the artisanal mining sector in South Africa. Although formalization is frequently proposed as a solution to the sector's challenges, the process is far from simple. The forthcoming chapter will explore the complex process of formalizing the artisanal mining sector.

References

Action Aid. (2019). *Regularising informal artisanal mining in South Africa: An evidence-based report*. Action Aid.

African Mining. (2024). *Safety in SA's mining sector – current trends, challenges and strategic opportunities*. Accessed November 17, 2024, from https://www.african-mining.co.za/2024/08/01/safety-in-sas-mining-sector-current-trends-challenges-and-strategic-opportunities/#:~:text=Current%20safety%20statistics&text=There%20were%20270%20fatalities%20in,20%20fatalities%20in%202023%20respectively

Andrew, J. S., & Hilson, G. M. (2003). Land use disputes between small- and large-scale miners: Improving conflict management. In G. M. Hilson (Ed.), *The socio-economic impacts of artisanal and small-scale mining in developing countries* (pp. 25–42). Balkema Publishers.

Business & Human Rights Resource Centre. (2024). *S. Africa: Research finds acid mine drainage has contaminated water sources, putting people's lives at risk*. Business & Human Rights Resource Centre. Accessed October 12, 2024, from https://www.business-humanrights.org/en/latest-news/s-africa-research-finds-acid-mine-drainage-has-contaminated-water-sources-putting-peoples-lives-at-risk

Buxton, A. (2013). *Responding to the challenge of artisanal and small-scale mining. How can knowledge networks help?* Accessed August 15, 2015, from http://pubs.iied.org/16532IIED.html

Centre for Environmental Rights. (2016, May). *Zero hour: Poor governance of mining and the violation of environmental rights in Mpumalanga*. Centre for Environmental Rights. Accessed October 12, 2024, from https://cer.org.za/wp-content/uploads/2016/06/Zero-Hour-May-2016.pdf

Chupezi, T. J., Ingram, V., & Schure, J. (2009). *Impacts of artisanal gold and diamond on livelihoods and the environment in the Sangha Tri-National Park Landscape*. Accessed March 1, 2016, from https://www.cifor.org/publications/pdf_files/Books/BChupezi0901.pdf

De Haan, J., Dales, K., & McQuilken, J. (2020). *Mapping artisanal and small-scale mining to the sustainable development goals. Minerals, Materials and Society program in partnership with Pact*. Accessed January 20, 2022, from https://www.pactworld.org/library/mapping-artisanal-and-small-scale-mining-sustainable-development-goals

Deberdt, R. (2021). The Democratic Republic of the Congo (DRC)'s response to artisanal cobalt mining: The Entreprise Générale du Cobalt (EGC). *The Extractive Industries and Society, 8*(4), 101013. https://doi.org/10.1016/j.exis.2021.101013

Department of Mineral Resources (DMR). (2015). Public Labour Annual Report. Pretoria: DMR.

Environmental Law Institute. (2014). Artisanal and small-scale gold mining in Nigeria. recommendations to address Mercury and Lead Exposure. http://www.eli.org/sites/default/files/eli-pubs/nigeria-asgm-assessment-final-report.pdf. Accessed 1 March 2016.

Hagos, G., Sisay, W., Alem, Z., Niguse, G., & Mekonen, A. (2016). Participation on traditional gold mining and its impact on natural resources, the case of Asgede Tsimbla, Tigray, Northern Ethiopia. *Journal of Earth Sciences and Geotechnical Engineering, 6*(1), 89–89. http://www.scienpress.com/Upload/GEO/Vol%206_1_6.pdf

Hosken, G. (2017). *We know it's dangerous, but the money is good – South Africa's Zama Zama mining explosion*. Accessed June 12, 2017, from http://www.timeslive.co.za/local/2017/05/21/WATCH-%E2%80%98We-know-it%E2%80%99s-dangerous%E2%80%9A-but-the-money-is-good%E2%80%99-South-Africas-illegal-mining-explosion

Human Rights Watch. (2013). *Toxic toil: Child labour and mercury exposure in Tanzania's small-scale gold mines*. Accessed April 2, 2019, from https://www.hrw.org/sites/default/files/reports/tanzania0813_ForUpload_0.pdf

International Labour Organisation (ILO). (2011). *Children in hazardous work*. Accessed March 1, 2016, from http://www.ilo.org/wcmsp5/groups/public/@dgreports/@dcomm/@publ/documents/publication/wcms_155428.pdf

International Labour Organization. (2019). *Child labour in mining and global supply chains*. https://www.ilo.org/wcmsp5/groups/public/%2D%2D-asia/%2D%2D-ro-bangkok/%2D%2D-ilo-manila/documents/publication/wcms_720743.pdf

Jinnah. (2016, January 25). *Between a rock and a hard place: Informal artisanal gold mining in Joh*. Africa at LSE. https://blogs.lse.ac.uk/africaatlse/2016/01/26/between-a-rock-and-a-hard-place-informal-artisanal-gold-mining-in-johannesburg/#comments

Mafata, M. (2024, May 23). South Africa: Gold mine pollution is poisoning Soweto's water and soil – study finds food gardens are at risk. *The Conversation*. Accessed October 12, 2024, from https://theconversation.com/south-africa-gold-mine-pollution-is-poisoning-sowetos-water-and-soil-study-finds-food-gardens-are-at-risk-229775

Martin. (2019). Uncovered: The dark world of the Zama Zamas. *ENACT Africa*. https://enactafrica.org/research/policy-briefs/uncovered-the-dark-world-of-the-zama-zamas

Mhangara, P., Tsoeleng, L. T., & Mapurisa, W. (2020). Monitoring the development of artisanal mines in South Africa. *Journal of the South African Institute of Mining and Metallurgy, 120*(4). https://doi.org/10.17159/2411-9717/938/2020

Minerals Council South Africa. (2023). *Illegal mining*. Accessed November 2, 2024, from https://www.mineralscouncil.org.za/work/illegal-mining

Mujere, N., & Isidiro, M. (2016). *Impacts of artisanal and small-scale gold mining on water quality in Mozambique and Zimbabwe*. https://pdfs.semanticscholar.org/0ae3/4a80b260c8eb4c55b375bf972172155f8166.pdf

Munnik, V. (2010). *The social and environmental consequences of coal mining in South Africa: A case study*. Accessed March 20, 2020, from http://www.bothends.org/uploaded_files/uploadlibraryitem/1case_study_South_Africa_updated.pdf

Njini, F. (2018, December 10). Gold industry enters final phase of slow death. *Business LIVE*. https://www.businesslive.co.za/bd/business-and-economy/2018-12-10-gold-industry-enters-final-phase-of-slow-death/

O'Driscoll, D. (2017). *Overview of child labour in the artisanal and small-scale mining sector in Asia and Africa*. Accessed May 1, 2023, from https://assets.publishing.service.gov.uk/media/5a5f34feed915d7dfb57d02f/209-213-Child-labour-in-mining.pdf

Ojewale, O. (2021, October 18). Child miners: the dark side of the DRC's coltan wealth can be countered by legitimate mineral certification. *Daily Maverick*. Accessed March 20, 2020, from https://www.dailymaverick.co.za/article/2021-10-18-child-miners-the-dark-side-of-the-drcs-coltan-wealth-can-be-countered-by-legitimate-mineral-certification/

Osumanu, I. K. (2020). Small-scale mining and livelihood dynamics in north-eastern Ghana: Sustaining rural livelihoods in a changing environment. *Progress in Development Studies, 20*(3), 208–222. https://doi.org/10.1177/1464993420934223

Paton, A. (1948). *Cry, the beloved country*. Jonathan Cape.
Persaud, A., & Telmer, K. (2015). *Developing baseline estimates of mercury use in artisanal and small-scale gold mining communities: A practical guide (Version 1.0)*. Artisanal Gold Council. Accessed January 15, 2019, from http://www.briloon.org/uploads/BRI_Documents/Tropical_Program/ASGM/AGC%20Inventory%20Guide_v1_Oct2015.pdf
Rajee, M., Obiri, S., Green, A., Long, R., Cobbina, S. J., Nartey, V., Buck, D., Antwi, E., & Basu, N. (2015). Integrated assessment of artisanal and small-scale gold mining in Ghana—Part 2: Natural sciences review. *International Journal of Environmental Research and Public Health, 12*, 8971–9011. https://doi.org/10.3390/ijerph120808971
Richard, M., Moher, P., & Telmer, K. (2014). *Health issues in artisanal and small-scale gold mining: Training for health professionals (Version 1.0)*. Artisanal Gold Council. Accessed March 1, 2016, from http://www.unep.org/chemicalsandwaste/Portals/9/Mercury/Documents/ASGM/Health_Training_Oct2014_version1.0_eng_lowQ.pdf
Singh, O., & Makhetha, M. (2023). *Bloody Zama Zama turf war threatens lives and livelihoods at landfill*. Accessed July 1, 2023, from Bloody zama zama turf war threatens lives and livelihoods at landfill (timeslive.co.za)
Snell, L. (2019). Death and destruction in South Africa's city of gold. *Reuters*. Accessed March 20, 2021, from https://www.reuters.com/article/us-safrica-mining-idUSKCN11J1M6
South African Human Rights Commission (SAHRC). (2018). *National hearing on the underlying socio-economic challenges of mining-affected communities in South Africa*. Accessed November 17, 2024, from https://www.sahrc.org.za/index.php/publications/investigative-hearing-report-on-the-underlying-socio-economic-challenges-of-mining-affected-communities-in-south-africa-august-2018/viewdocument/100
Stewart, P., Bezuidenhout, A., & Bischoff, C. (2019). Safety and health before and after Marikana: Subcontracting, illegal mining and trade union rivalry in the South African mining industry. *Review of African Political Economy, 47*(163), 27–44. https://doi.org/10.1080/03056244.2019.1679103
Thornton, R. J. (2014). Zamazama, "illegal" artisanal miners, misrepresented by the South African Press and Government. *The Extractive Industries and Society, 1*(2), 127–129. https://doi.org/10.1016/j.exis.2014.06.003
United Nations Economic Commission for Africa. (2011). *Minerals and Africa's development*. Accessed November 19, 2021, from https://archive.uneca.org/publications/minerals-and-africas-development
United Nations Environmental Programme (UNEP). (2012). *A practical guide: Reducing mercury use in artisanal and small-scale gold mining*. Accessed July 3, 2019, from https://wedocs.unep.org/bitstream/handle/20.500.11822/11524/reducing_mercury_artisanal_gold_mining.pdf?sequence=1&isAllowed=y
United States Environmental Protection Agency. (2024). *Mercury: Health effects of exposures to mercury*. Accessed October 12, 2024, from https://www.epa.gov/mercury/health-effects-exposures-mercury
World Health Organisation (WHO). (2016). Artisanal and small-scale gold mining and health. 9789241510271-eng.pdf (who.int). Accessed 14 January 2019.

Chapter 7
Between a Rock and a Hard Place with the Complexities Around Formalization

Our focus has to be on changing reality, not changing laws.

—Sheryl Wudunn (Kristof & WuDunn, 2010)

Introduction

Several social and environmental issues are linked to the artisanal and small-scale mining (ASM) sector. Formalizing the sector is regarded as one way to address these issues and all the ills and challenges. Still, this is not always so easy, and on many occasions, formalization efforts have failed.

This chapter looks at what formalization is, in general, and what formalization entails when we talk about formalizing the ASM sector.

A closer look at the formalization processes in two other African countries, namely Ghana and the Democratic Republic of the Congo (DRC), is provided.

Examining the implementation of formalization practices in these two countries reveals some pitfalls and underscores the difficulties of formalizing a sector like artisanal mining. Following this, an overview of the South African government's approach to artisanal mining to date is presented.

The concluding section of this chapter examines the implications of formalization practices for women engaged in artisanal mining, an aspect frequently overshadowed in the broader discussions surrounding formalization.

What Is Formalization?

Overall, formalization can be understood in different ways. For one, formalization can imply removing people from the informal sector and moving them to the formal sector (Chen, 2018). However, as it stands, the formal economy cannot create enough jobs for the unemployed, let alone those in the informal economy (Chen, 2018). Another way to interpret formalization is that it entails the tax registration of informal enterprises. The issue with such an approach is that many informal

V. Bester, *The Untold Story of Zama Zama Miners in South Africa*, Clinical Sociology: Research and Practice, https://doi.org/10.1007/978-3-031-82783-9_7

enterprises already pay license fees, operating fees, or market rents and already pay value-added tax on purchased goods (Chen, 2018). A third approach could involve focusing on the sector's regulation by forcing the informal sector to comply with licensing and business zone requirements (Fourie, 2018). This entails the authorization and control of business settings, which could be damaging if the relevant authority enforces compliance to gain revenue (Fourie, 2018).

The International Labour Organisation (ILO) developed and adopted Recommendation 204 in 2015 to respond more developmentally to formalization. This recommendation focuses on transitioning from the informal economy to the formal economy. Also, Recommendation 204 calls for the government to create enabling environments for informal employees and enterprises to exercise their right to organize and bargain. As part of Recommendation 204, the ILO also endorses a social dialogue between participants in the informal sector and those in government (Chen, 2018). Furthermore, scholars suggest that an alternative tactic to formalization is to adopt a developmental approach.

A developmental approach to formalization entails two suggestions (Fourie, 2018). Firstly, formalization should not be regarded as the end goal or a policy accomplishment. Instead, formalization should be considered a means to aid the quest for better livelihoods and more substantial and self-sufficient informal initiatives. Secondly, such a formalization approach may require a component of registration and licensing for an informal enterprise to receive support. Support can include, but is not limited to, obtaining access to appropriate business premises, access to essential services such as water and electricity, and receiving bookkeeping training. These supporting elements may vary and depend on the type of sector that needs support and may be seen as steps in the formalization process (Fourie, 2018). This introduces a much more flexible approach toward formalization that enables the development of specific informal sectors. Several degrees of formalization should co-occur, along with the range of informal enterprises (Fourie, 2018).

Formalization is a general response to address the ASM sector. International organizations and governments often support and consider formalization the primary intervention to address this sector's social and environmental issues (Siwale & Siwale, 2017). The laws of a country determine the legal, illegal, informal, or formal status of artisanal mining; therefore, governments worldwide are regarded as a primary force in transforming the ASM sector (IIED, 2015).

Formalizing ASM: A Means for Development, Not the End

The theoretical assumption of ASM formalization is that economic development will occur when formal property rights are intact. To support this standpoint, it has been argued that when land titles are formal, a realization around notions of order and security is formed (Geenen, 2012). According to this view, formalization will propel owners to respect other people's titles, and at the same time, physical and human security will increase.

For example, artisanal miners are often victims of extortion and excessive bribes, or governments or mining companies threaten them with violent evictions. In the artisanal miners' case, Siegel and Veiga (2009) argue that formal titles are "*...the basis of a miner's access to legal redress when rights are violated. Therefore, formalization will give the artisanal miners a 'voice' in decision-making.*"

Similarly, the work of Hernando de Soto (and the legalist school of thought) argues that there is a strong link between property rights and economic development. de Soto contends that granting people formal property rights is a fundamental condition for financial growth (Geenen, 2012). de Soto maintains that formal titles, in terms of property, would create incentives for investments and resource development. Formal titles will also grant access to credit, and at the same time, formal titles can be transformed into standardized tools of exchange and capital (Geenen, 2012). According to de Soto's argument, this will pave the way for individuals to become active members of the market economy. In de Soto's view, formal titles will increase wealth and economic growth, but these formal titles will not occur naturally. They should, it is suggested, be created through legal reforms, such as formalization (Geenen, 2012).

The issue of access to land and property is one of the main stumbling blocks for artisanal miners. It is easy to see why the legalist school of thought follows de Soto's argument regarding the formalization of ASM. This discourse has been highly prominent in policy circles (Gilbert, 2002; Geenen, 2012). As de Soto argues, formalization is ultimately a clash between the mainstream economy, namely the formal economy and the informal, extra-legal economy (Siegel & Veiga, 2009; Geenen, 2012).

Regarding the informal economy, property rights consist of customary rights and possessory relationships. According to de Soto's argument, people experiencing poverty in the informal economy lack formal titles and cannot transform their formal titles and assets into security and credit. Therefore, their assets are tied up in what de Soto terms "*dead capital*" (Siegel & Veiga, 2009). The impoverished need, so the argument goes, secure tenure titles that the government's legislative framework should safeguard and enforce (Geenen, 2012).

The relevance of de Soto's argument becomes more apparent when considering that most of the world's workforce is part of the informal economy. The ILO estimates that over 2 billion of the world's employed population (aged 15 and older) work informally (ILO, 2018). This means that over 60% of the global workforce is part of the informal economy.

Most criticism of de Soto's ideas focuses on the effectiveness of his recommendations (Hilson et al., 2017). Critics have remarked that creating wealth is not as straightforward as granting tenure rights. Even when permits and titles are recognized (see Kingwill et al., 2006; Nyamu-Musembi, 2006; Hilson et al., 2017). Development does not follow such a linear course. In terms of ASM, Geenen (2012) highlights four key criticisms of de Soto's and the legalist school of thought, which are as follows:

1. Firstly, Geenen (2012) states that the official registration of mining activities has been considered a means to address ASM problems too often. However, these actions are tremendously bureaucratic, inappropriate, and incompatible with realities (Hilson, 2009). Siegal and Veiga (2009) cautioned that formalization is a process, not a product. Moreover, the most elaborate policies to formalize mining activities fail if a government lacks the will to implement these plans if artisanal miners perceive licensing as a threat, or if the miners cannot afford the costs of joining the legal economy, these efforts fall flat.
2. Secondly, the formalization discourse assumes everyone will eventually benefit from considerable property rights. However, it has been debated that not all artisanal miners want formal property rights. Some miners do not consider their illegal status an issue if they have alleged tenure security (Geenen, 2012; Gilbert, 2002). For example, customary titles can be just as valid, sound, and sometimes even more legitimate (Gilbert, 2002). Furthermore, what property rights are preferred depends on the context and nature of the mining activities.
3. Thirdly, the standard formalization canon does not consider the informal economy's diversity and complexity. Geenen (2012) rightly states that the economic activity of mineral exploitation in the DRC is closely interwoven with a whole web of social relations. In the South African context, the same can be said. For example, some women work as Zama Zamas and use their earnings to start a small business such as an informal restaurant called *shisa nyama. Shisa nyamas* are the main eating places for many Zama Zamas and people living in informal settlements (Gauteng City-Region Observatory, 2018). The same observation was made during my field research when I witnessed a broader informal economy around many artisanal mining sites, with vendors selling various goods from gumboots to hand tools to food.
4. Another crucial point is how deeply interwoven artisanal mining activities are with mineral trade routes. This includes smuggling routes and, in the South African context, the role of syndicates. The fact is that any formalization strategy should consider the artisanal sector's dependency on these informal and illegal mineral trade routes, power relationships, and the complex web of interactions between the miners and the traders. This is also the case in South Africa. The role of the syndicates cannot be ignored in a formalization strategy. Any means to formalize the sector should be cognizant that a network of syndicates is likely part of the artisanal miners' trade.
5. Lastly, some empirical studies from developing countries have confirmed that formal titling risks reinforce inequalities by benefiting the rich and middle class rather than the poor (Gilbert, 2002; Plateau, 2015). With mining titles, governments tend to open their domestic markets to foreign investors and give preference to large-scale mining (LSM) companies. Selections are often made explicit in mining laws, tax holidays, or protracted titles, thereby disadvantaging ASM (Hilson & Potter, 2005; Clausen, 2011). Alternatively, as noted by Maconachie and Hilson (2011), the emphasis on large-scale gold mining development has frequently hindered the achievement of mining sector reform, specifically the integration of small-scale operators into the legal framework.

The points mentioned above are vital when considering a formalization approach. The formalization blueprint prescribed by de Soto is nowhere close to being bulletproof. Still, as Hilson et al. (2017) point out, de Soto's thoughts are significant when explaining how governments potentially create informality, especially when looking at ASM in sub-Saharan Africa.

To support his ideas, de Soto refers to Peru by explaining how it took up to 289 days to open a business in Lima. Also, to secure a legal land title, 728 administrative steps were needed to complete the process (Hilson et al., 2017). This underlines that governments create stumbling blocks in a formalization process with cumbersome bureaucratic hurdles.

In sub-Saharan Africa, countless examples show how governments create informality through bureaucracy and costs. This is comparable to de Soto's Peru description of how policy systems were driving the growth of the country's extra-legal economy (Hilson et al., 2017). Furthermore, in Ghana, for instance, scholars have highlighted that over the past 15 years, artisanal miners have repeatedly complained to government officials about the fees for a license and the then requirements for an environmental permit, which many artisanal miners claim have kept them in the informal economy (e.g., Hilson & Potter, 2005; Tschakert & Singha, 2007; Hilson et al., 2017).

As can be seen, the matter of formalization is rather complicated. For this reason, it is suggested that a good starting point for an effective formalization strategy is to clarify what formalization means in ASM (Hilson, 2020). The following United Nations (UN) definition encapsulates the donor position on formalizing the ASM sector. With a focus on gold, the UN states that formalization:

> …seeks to integrate [artisanal and small-scale gold mining] ASGM into the formal economy. The process of formalization includes the development or adaptation of mining (and other) laws or policies to address the challenges of ASGM. A well-designed formalization process generates the enabling conditions for accountability within the sector so that it can ultimately be integrated into the formal economy. Formalization can only be successfully achieved if programs and public policy deal with the different dimensions of ASGM activities simultaneously and in an integrated way. Legalization is just one dimension of the process of formalization.

According to the United Nations' definition, legal and policy frameworks are vital for formalization (Hilson, 2020). However, more importantly, formalization should be a process that can integrate the ASM sectors into the formal sector. Policy changes should serve as a first step in the formalization process but should not be the last. Formalization in ASM usually entails legislation amendments, which are much needed and serve as a good starting point for formalization.

However, the issue with granting legal titles is that governments assume that formalization ends when legal titles are granted rather than viewing the process as ongoing (Dzimunya et al., 2018). Furthermore, other donor documents, such as those from the United Nations Institute for Training and Research (UNITAR) and the UN Environment, share similar ideas on this matter (Hilson, 2020). For example, it is recommended that licensing schemes and policy frameworks should serve as the foundation for formalization and be regarded as a platform to launch support services for individual operators.

The World Bank's 2020 State of the ASM Sector report mentions "*holistic formalization*" (World Bank, 2020; Robles et al., 2022). In the World Bank's report, explicit attention is paid to labor. This aligns with Sustainable Development Goal 8 (decent work and economic growth). According to the World Bank's 2020 report, if effectively harnessed, ASM has the potential to provide more inclusive and sustainable economic growth, productive employment, and, most importantly, decent work for tens of millions of people worldwide (World Bank, 2020; Robles et al., 2022). However, workers do not always prefer to automatically reject flexible labor arrangements or revenue-sharing models (typical of the informal economy) in favor of fixed salaries (Robles et al., 2022).

Many sub-Saharan African countries have tried to formalize their ASM sectors by amending or introducing legislation (Hilson et al., 2017). Nevertheless, operations in sub-Saharan Africa are still primarily unlicensed and limited to informal spaces despite formalization efforts (Hilson et al., 2017). In agreement with this, Siwale and Siwale (2017) stated that experiences from several of the region's countries prove that formalization does not have the desired outcome. Scholars have argued that many of these countries lack specific ASM legislation that explicitly addresses the nuances of ASM (Collins & Lawson, 2014; Mcquilken & Hilson, 2016). In addition, the IIED's research stated that governments' lack of knowledge, capacity, and political will are hurdles to overcome when improving the ASM sector (IIED, 2015). Such capacity and technical shortages make formalization less straightforward in any government's quest to address ASM.

Some companies provide support in this regard to address this challenge. For example, the national ASM formalization strategy in Mozambique requires government engagement. Levin Sources and the Alliance for Responsible Mining (ARM), with Genesis Lda, integrated capacity-building efforts in the World Bank-funded Mining and Gas Technical Assistance Project (MAGTAP). This approach shifted from traditional one-off training to a comprehensive program of ongoing stakeholder engagement and learning, focusing on formalizing ASM pilot areas and developing extension services recommendations. The project team collaborated closely with the Ministry of Energy and Mines in Mozambique and provincial and district authorities to deliver effective capacity building (Levin Sources, n.d.).

To discuss the critique of formalization further, country-specific formalization initiatives will now be evaluated. With ASM occurring in over 80 countries, the literature shows abundant formalization strategies and initiatives developed and implemented over the past decades. Ghana is included since it is regarded as a pioneer in formalizing the ASM sector (Hruschka, 2015). Also, regarding its formalization approaches to ASM, the country shows many similarities to the current situation around ASM in South Africa.

The Case of Artisanal and Small-Scale Mining Formalization in Ghana

Ghana is Africa's largest gold producer and ranked sixth globally in 2023 (Boafo, 2024). In Ghana, approximately 1 million artisanal and small-scale miners extract gold, making it the primary mineral for this workforce. As of 2023, the artisanal and small-scale gold mining (ASGM) sector contributed about 35% to the nation's total gold production (Adranyi et al., 2024).

The formalization of ASM in sub-Saharan Africa started in Ghana, where it implemented its Small-Scale Gold Mining Law in 1989 (Collins & Lawson, 2014). In short, the law allows mining licenses to be issued to small-scale miners if they fulfill the requirements. Initially, to be given a small-scale mining license, an environmental assessment by the government's environmental protection agency had to be done (Debrah et al., 2014). Since 2016, this is no longer a prerequisite (Mcquilken & Hilson, 2016).

However, the process remains a demanding task for aspiring small-scale miners. A mining license for a concession of 0.4–2 ha is subject to government review and approval every 3 years, and concessions between 2 and 10 ha must be reviewed every 5 years (Mcquilken & Hilson, 2016). Applicants must submit site plans of the area, complete application forms and copies of such, and pay a fee to the Minerals Commission (Debrah et al., 2014).

Several projects have been sponsored and implemented to support Ghana's ASM sector. For example, the Small-Scale Mining Project (SSMP) was introduced in 1989 to handle relevant policy-related matters in the sector (Hruschka, 2015). The project aimed to issue licenses and implement applicable industry policies. Furthermore, in 1991, the Small-Scale Mining Department (SSMD) was established to address industry policy and regulatory issues better. The SSMD was strengthened during the mid-90s by the World Bank's financial support, which contributed to developing a Mining Sector Development and Environmental project (Hruschka, 2015). The project's primary aims were twofold: to bolster the capability of mining sector institutions to effectively promote and regulate investments in mining while prioritizing environmental considerations and to advance methods to enhance small-scale mining operations (Hruschka, 2015).

Despite these initiatives to support the small-scale mining sector, the miners still find it extremely challenging to meet the demands of a mining license. These formalization strategies from the government are having little success, with numerous recent news reports on Ghana's battle with illegal mining (locally known as *galamsey*) (Kumah, 2021). Regardless of the formalization strategies, most miners still operate informally or illegally without any mining license.

The matter of land has been highlighted as a critical point in the formalization process. For that reason, the Ghanaian government has allocated land for small-scale mining, but the exercise has been ineffective (Hilson, 2017). Instead of identifying areas viable for mining, the government has waited for mining companies to abandon their concessions, resulting in some concessions not having viable quantities of gold available anymore (Hilson, 2017). The same proposed initiative is recorded in the South African ASM policy, which is discussed in more detail in the latter part of this chapter.

Recently, an influx of Chinese miners has been recorded in Ghana. Since the early 2000s, about 50,000 Chinese gold miners have migrated to Ghana for small-scale mining (Boafo et al., 2019). The hike in gold prices from 2008 onwards led to a gold rush, which caused the influx of significant numbers of foreign miners, mainly working illegally, given that the sector is reserved for Ghanaian citizens (Crawford et al., 2015).

By 2013, Chinese involvement in informal gold mining in Ghana had grown to such proportions that the government of Ghana felt forced to act since the media covered stories of these Chinese miners in a very hostile manner (Crawford et al., 2015). In mid-May 2013, the then and current president of Ghana, Mr. John Mahama, established a high-level Inter-Ministerial Task Force to combat illegal small-scale mining. While Mr. Mahama was cautious about including all illegal mining activities of Ghanaians and non-Ghanaians, it was clear that the measures were primarily aimed at foreign miners (Crawford et al., 2015). The Task Force was a military operation undertaken by Ghana's police and army to *"flush out"* the illegal miners. This strategy mirrors the South African government's current way of handling the Zama Zamas situation. In mid-2013, the measures led to the arrest and deportation of many foreign nationals, of which the bulk were Chinese miners (Crawford et al., 2015). Despite this so-called *clean up galamsey* initiative, illegal mining continues in Ghana.

Hilson et al. (2017) rightly remark that the real question should uncover why people are moving into this. Efforts should be directed at what drives the sector. For example, questions should be about the following: *"How have the problems which are being condemned come about? As opposed to publicly referring to galamsey as 'a menace'"* (Hilson, 2017).

This is a critical comment for consideration in the South African context. There exists a sad tendency toward pocket incidences of xenophobia in the country. Any strategy or measure targeting foreigners, however subtle it may be, should not be considered at all or be part of the formalization of the sector. A sad truth is that the South African mainstream media, as in Ghana, is also creating a picture that illegal immigrants cause Zama Zama mining in the country. Therefore, one must consider the broader implications of any formalization strategy that ignores foreigners. Having a formalization strategy that excludes foreigners has the potential to spur xenophobic attacks. Laws might operate in clinical ways, but people do not. Our social lives are layered like tectonic plates; any ripple or change on one surface or below the surface affects the wider society.

The parallels between ASM in Ghana and ASM in South Africa are striking. Both countries included provisions for small-scale mining in their key mining legislation without mentioning artisanal mining. Neither the Mineral and Petroleum Resources Development Act (MPRDA) of 2002 nor Ghana's Mineral and Mining Act of 2006 recognizes the diverse requirements of the artisanal and even the small-scale miner (Debrah et al., 2014).

Both key mining legislation pieces speak to only one part of existing ASM actions: small-scale mining. This leaves the artisanal miners who function at the spectrum's lower end excluded from legislation (Debrah et al., 2014). Consequently, the governmental responses to both artisanal mining activities in Ghana and South Africa similarly address the matter by deploying security forces.

Turning to another example of ASM formalization, the Democratic Republic of the Congo (DRC) situation will be considered. The DRC is one of the few countries whose legislation explicitly recognizes artisanal miners. It is, therefore, worthwhile to look at how the process and outcome of formalization are playing out in the DRC.

The Case of Artisanal and Small-Scale Mining Formalization in the DRC

Artisanal mining in the DRC was first recognized in its 2002 Mining Code by allowing artisanal exploitation zones. Requirements under the Mining Code stipulate that artisanal miners must apply annually for an artisanal mining card. Additionally, artisanal miners must form a cooperative to access the artisanal mining zones (Singo & Seguin, 2018). These miners must also meet the requirements and regulations regarding health, safety, and the environment (Singo & Seguin, 2018).

Artisanal mining zones were again recognized in the country's updated Mining Code, adopted in March 2018. These designated zones are the primary practice for providing artisanal miners access to land (Singo & Seguin, 2018). Zones are allocated in areas where it has become unfeasible for industrial mining to continue to operate (Singo & Seguin, 2018). However, the Mining Code tried to revive the LSM sector by attracting foreign investment (Karaki, 2018). This resulted in large concession areas being given to mining companies, even though artisanal miners were sometimes present on the concessions (Karaki, 2018). Furthermore, the law allows closing any artisanal exploitation zone if new deposits are feasible for LSM (Geenen & Claessens, 2013). Should a company or individual obtain mineral rights in an area where artisanal miners are mining, the artisanal miners are required to stop their operations and move away (International Conference on the Great Lakes Region (ICGLR), 2017). According to the country's Mining Code, artisanal miners are usually given 2 months' notice to vacate a concession should an LSM company apply to mine the same land (Singo & Seguin, 2018).

Consequently, when artisanal miners are removed from the land, they continue outside the designated artisanal exploitation zones. Resulting in the miners operating outside any legal framework, which fuels conflict between the artisanal mining sector and mining companies (Karaki, 2018). Furthermore, the miners have limited capital and geological knowledge, which has led to some of these designated artisanal mining zones not being mined (Singo & Seguin, 2018). In June and July 2019, security forces evicted artisanal miners who mined on two of the country's largest industrial mining sites in the Haut-Katanga and Lualaba provinces. Besides deaths and injuries, the expulsions caused more than 10,000 artisanal miners to lose their only income source (International Crisis Group, 2020).

Moreover, the requirements to obtain an artisanal mining card (a requirement under the country's Mining Code) signify barriers for artisanal miners to mine legally (Karaki, 2018).

One can then correctly argue that although artisanal mining is recognized explicitly in the Congolese Mining Law, formalization still falls short by not providing supportive initiatives for the miners. Ultimately, the artisanal miners operate without any artisanal mining card or license. Additionally, the recognition and sanction of artisanal mining zones have been unsatisfactory; the result is that artisanal miners mine in areas not lawfully zoned off (Singo & Seguin, 2018). The DRC experience indicates that solid implementation of legislation is necessary, coupled with realistic regulatory expectations and ongoing support.

The following section looks at the South African government's journey in terms of ASM formalization.

The South African Government and the Artisanal and Small-Scale Mining Sector

The mining history in South Africa is roughly clustered into three mineral revolutions. The gold rush periods can be divided into the Pilgrims Rest Gold Rush (1873–1881) and the Witwatersrand Gold Rush, which started in 1886 when gold was discovered in the world's largest gold basin. In its early days, the Witwatersrand basin was beset with many artisanal and small-scale miners, all in search of a piece of gold. Nevertheless, in less than 5 years, a viable and profitable gold mining industry was established on the Witwatersrand (Davenport, 2013). By 1890, major European investment houses and banks pioneered the establishment of large mining houses that took control of the South African mining industry (Davenport, 2013).

The same trajectory occurred in the diamond sector, resulting in another, the third, mineral revolution. After discovering the first diamond in 1866, diamond diggers beset the diamond fields of the then-Griqualand (the Northern Cape Province), South Africa (Meredith, 2007). In both the gold and diamond rushes, it was not long before mining houses took control of the mineral riches of South Africa by pushing these early-day artisanal miners off the land.

Consequently, since the early 1900s, LSM has controlled the extractive landscape of South Africa. With the government boosting LSM, little room was left to incorporate small-scale mining, not to mention artisanal mining, into mining legislation. This one-sided approach from the government unfairly favored the mining houses, which prosperous investors dominated. Subsequently, little effort was made to include artisanal and small-scale miners in the country's economic development, thus creating an equity gap. During the apartheid era, this equity gap was further deepened, and no opportunities existed for artisanal and small-scale miners to benefit from the mineral richness of South Africa (Ledwaba, 2017). As seen in Chaps. 2 and 3, the pre and apartheid eras created deep pockets of socioeconomic exclusion along race and class lines.

After 1994, with the fall of apartheid, legislation was implemented to foster ASM in South Africa. The Reconstruction and Development Programme (RDP) of 1994 recognized the potential of ASM to widen access to the mining sector (UBC and CIRDA, 2019). The RDP encouraged the development of the small-scale mining sector by providing financial and technical aid and access to mineral rights (UBC and CIRDI, 2019). Another significant publication for ASM development was the 1998 White Paper on Minerals and Mining Policy of South Africa. In the White Paper, the government declared its intent to:

> Encourage and facilitate the sustainable development of small-scale mining to ensure the optimal exploitation of small mineral deposits and enable this sector to make a positive contribution to the national, provincial, and local economy. (UBC and Canadian International Resources and Development Institute (CIRDI), 2019)

Furthermore, the White Paper rightly pointed out that small-scale and artisanal mining need to be defined for clarity. Also, the White Paper stated that small-scale mining ranges from tiny operations that provide subsistence living (for artisanal miners) to *"junior"* companies for which subsistence living is not the prime motivator. Moreover, the White Paper recognized the role that small-scale mining could play in providing a channel for increased access to the mining industry but also identified several constraints to the realization of small-scale mining opportunities, including a lack of access to mineral rights, poor access to finance, a lack of management, marketing, and technical skills, poor access to significant mineral markets, and incoherent support structures to assist small-scale mining development. Several interventions were identified in the White Paper to address these constraints and thereby reinforce efforts to promote small-scale mining by:

- Improving access to geological, mining, environmental, and mineral market information.
- Increasing access to mineral rights and land.
- Streamlining regulations governing small-scale mining.
- Increasing the efficiency of administrative procedures.
- Promoting research and development for small-scale mining.
- Providing technical assistance and training to small-scale miners.
- Technology transfer.
- Reducing financial constraints to participation in small-scale mining and
- Ensuring an integrated approach to public and parastatal sector efforts to promote small-scale mining.

What is apparent is the oversight and rather indifferent attitude toward artisanal mining. This attitude is most visible in the text where the White Paper stated that:

> The interests of the country and the community demand that all forms of mining, whether large, small or artisanal, should be subject to the same requirements in respect of licensing, safety, health and the environment.

The White Paper continued by stating that:

> …the deleterious effects of artisanal or subsistence mining on the environment and on safety and health elsewhere in the world dictate the necessity for research in this area. Meanwhile, resources need to be employed by the State to control artisanal mining as effectively as possible.

The White Paper stated outright that artisanal mining, which has frequently caused severe environmental damage in other countries, should not be treated more leniently. According to the White Paper, uniform environmental management standards should be applied across mining operations of varying scales so that all mining is environmentally responsible. How the White Paper envisioned that artisanal miners should fulfill such ambitious tasks is unclear.

As mentioned earlier in the book (see Chap. 3), another critical piece of legislation that creates opportunities for historically disadvantaged South Africans is the MPRDA of 2002 (Moraka & Van Rensburg, 2015). The Act, developed and implemented to make room for ASM, ironically focuses on LSM and small-scale mining (SAHRC, 2015). To mine lawfully on a small-scale requires submitting an Environmental Management Plan to the Department of Mineral Resources and Energy (DMRE).

This Plan is a comprehensive document that entails, among other things, details such as the locality, a project description of the prospecting mine, and a detailed scientific account of the mine's possible impact on the environment. In addition, the application to become a small-scale miner is made online through the DMRE's portal called the South African Mineral Resources Administration System (SAMRAD). These requirements are not unreasonable, but they are only achievable for a specific group of people with the finances, technology, and infrastructure to meet them. SAMRAD creates barriers for applicants from remote areas with limited access to computers and the internet (UBC and Canadian International Resources and Development Institute (CIRDI), 2019). Unfortunately, this is the reality for many impoverished South Africans.

Considering the challenges outlined above, it becomes clear that the only option is to mine without a mining permit to earn a livelihood. The issue is that there is no other permit, right, or license for which a miner may apply to be recognized as a legal small-scale miner (Legal Resources Centre, 2016).

Two years after the MPRDA of 2002 was enacted, the then-DMR established the Directorate of Small-Scale Mining. The aim was to promote the development of the small-scale mining sector in South Africa. The Directorate aimed to provide support to aspiring small-scale miners in several ways, such as guidance in terms of the identification of mineral deposits, assistance in terms of environmental impact assessments (EIAs), advice in terms of legal arrangements, help with a mining feasibility study, and assistance with a market study (Department of Mineral Resources, 2017). However, these attempts to regulate the ASM sector and promote the small-scale sector have been largely unsuccessful (Ledwaba & Nhlengetwa, 2016).

A lot of research, and the accounts in this book, clearly show that artisanal miners still face challenges in accessing mineral rights, markets, capital, information, and appropriation technology and that there is an overall lack of institutional support (Hoadley & Limpitlaw, 2004; Ledwaba & Nhlengetwa, 2016; Ledwaba, 2017). Supportive of this, the South African Human Rights Commission reported in 2015 that there is no indication of the progress made by the then-DMR to regulate and support the small-scale mining sector (SAHRC, 2015).

In 2021, the South African government released the first draft of the ASM policy for public comment (DMRE, 2021). Subsequently, the government made the ASM policy available for implementation in March 2022 (DMRE, 2022). This marks the first step toward formalizing the artisanal mining sector in South Africa.

The ASM policy states that the minister should be empowered by legislation to reserve areas for the ASM industry. The policy also mentions that the Directorate is inadequate to support the ASM industry, and there exists a lack of cooperation and coordination between relevant governmental departments (DMRE, 2021). Collaboration, according to the policy, should be established with the Council of Geoscience (CGS), the Mine Health and Safety Council (MHSC), and the Department of Environment, Forestry, and Fisheries (DEFF). The CGS should provide geological data to support the process. Other institutions should facilitate relevant information and data for the necessary authorizations (such as Environmental Impact Assessment and Mine Health and Safety plans).

Governments ignore the implementation of comprehensive ASM policy frameworks and often start the formalization process by simplifying the strategy by re-categorizing the sector's activities by permits, levels of mechanization, and production (Bester, 2022). The same pitfall is noticeable in South Africa's ASM policy, which sets out monetary thresholds for differentiating artisanal mining with a maximum revenue of ZAR 1 million and small-scale mining with a maximum of ZAR 10 million. Also, the policy introduces *"a graduation provision,"* which allows artisanal miners to graduate to small-scale mining and for a small-scale miner to graduate to a medium, junior, or emerging miner based on production and investment thresholds. Even so, the policy does not provide details about the graduation provision and its intended operation (Adonisi-Kgame et al., 2022). Hilson and Maconachie (2020) explain that, at best, this type of formalization strategy puts the artisanal miners in a semi-formal state, with many artisanal miners that will still struggle to mobilize finance and mechanize their operations because of the rules they must adhere to.

This challenge is not novel and as mentioned earlier, research has shown that small-scale prospecting miners find the bureaucratic process and the exorbitant fees extremely challenging (Tschakert & Singha, 2007; Teschner, 2012; Afriyie et al., 2016; Hilson, 2017). The result is that most artisanal miners cannot meet these requirements due to a lack of resources. When they fail to do so, they are considered to be illegal miners.

Unfortunately, the same fate awaits any South African artisanal miner who cannot legally fulfill the requirements to mine. The ASM policy states that a transition period will allow all illegal miners to come forward and subject themselves to the formalization process. Also, the ASM policy stipulates that law enforcement agencies must step in if the illegal miners do not take advantage of the transition period—falling back, yet again, on policing the matter.

Furthermore, according to the ASM policy, ASM entities will be required to pay the government royalties and taxes (for example, Value Added Tax), subject to various incentives established by the National Treasury in favor of small businesses. The policy also envisions stringent reporting standards to track the progress of the ASM industry and mandates that the DMRE enforce monthly, quarterly, and annual reporting requirements for the ASM industry for production, employment statistics, and tax and royalties' payments. What the policy fails to mention is how support will be given to the artisanal miners to meet these stringent requirements.

One can argue that it is necessary to insist that artisanal miners should not mine without the mandatory mining license (Afriyie et al., 2016). After all, such compliance will ensure that the activities of these miners are effectively regulated. However, since some communities have limited livelihood options, stopping these miners is not simple. Therefore, preventing the miners from earning a living is more problematic and inhumane than allowing them to mine without a permit (Afriyie et al., 2016). Also, how can we ensure that placing a ban on these activities will be effective if communities lack access to sustainable livelihood sources?

Arrest or deportation will not prevent or deter this sector (SAHRC, 2015). Because using military forces to clamp down on illegal and artisanal mining has never been successful (Afriyie et al., 2016; Hilson, 2017).

Similarly, calls from civil society indicate that a different approach—other than criminalizing the sector—is needed. Not only has policing the sector failed thus far in South Africa and other parts of Africa, but it also suffocates a socioeconomic opportunity; however, if addressed and supported, the sector offers employment and growth opportunities to those who would not otherwise be able to access it.

What Formalization Means for Women Artisanal Miners

The position and experiences of women in artisanal mining were examined in Chap. 5. The chapter highlighted the diverse experiences of women compared to men and the variations among women themselves. Therefore, it is essential to integrate the issue of women in artisanal mining into the formalization process.

The formalization process is complex and, at the very least, should involve a process encompassing a more comprehensive approach than a mere legislative change. Since gender issues and challenges are prominent in the ASM sector, the gender element cannot be ignored during formalization.

Research has indicated that any empowerment and capacity-building exercise that attempts to connect with women working in the region's informal ASM spaces must be able to distinguish between the positives and negatives of this work, as well as understand the details of the circumstances that drove them to the sector in the first place (Hilson et al., 2017).

As a starting point for any formalization process for women ASM operators in sub-Saharan Africa, policy designers are advised to follow the recommendations of Bashirwa et al. (2014), who conclude that *"..given the lack of alternative livelihoods in the Kivus, we believe that it makes more sense to address the predicament of women in eastern DRC's artisanal mining areas by taking the latter's agency into account"* (Hilson et al., 2017). It's crucial to consider the agency of women in the artisanal mining sectors when addressing their challenges.

An overarching trend identified in the research is that formalization initiatives, such as encouraging miners' associations or cooperatives, have been characterized by increased elite dominance.

So far, government regulatory responses to formalize ASM and improve sector conditions have been inconsistent in their attention to women and the gendered division of labor. Some governments are becoming more aware of and proactive in addressing the gender dynamics of ASM in various ways. For example, the African Union's Africa Mining Vision emphasizes incorporating gender equity into mining policies and legislation. Similarly, the Minamata Convention on Mercury requires state parties to include strategies to prevent the exposure of vulnerable populations, particularly children and women and pregnant women, to mercury, used in artisanal and small-scale gold mining, in their national action plans (NAPs). Despite international efforts, women remain largely absent from worldwide ASM legislation. ASM laws in countries such as Ghana, the Philippines, and Peru, for example, make no mention of women's specific roles in and experiences within ASM (Tobalagba & Vijeyarasa, 2020).

As for South Africa's ASM policy, the document is vague on the role of women and any means to involve them in the formalization process. The policy refers to women once under the section government policy proposals, subsection (c) licensing criteria. It vaguely states, *"Prioritize the interest of women and vulnerable groups involved in the ASM industry."* Yet, the policy fails to mention how this should be done in reality.

A gendered response to ASM could result in women having access to gender-responsive services and being protected from the specific vulnerabilities and challenges associated with their roles in ASM. Provided they form a core part of any formalization process, over time, access to increased training and leadership opportunities may facilitate women's access to microcredits, mining licenses, and business development, as well as their enjoyment of improved health and safety practices. Gender-conscious legislation could increase women's opportunities to actively participate in decision-making on public policy issues affecting women at community levels (Tobalagba & Vijeyarasa, 2020).

Simultaneously, gender-disaggregated data collection in ASM will be critical for better understanding women's experiences in the sector and the effectiveness of legislative responses. A top-down regulatory approach that advocates for gender-responsive initiatives will be insufficient. What is needed are the voices of the women themselves to ensure gender equality in the sector. The law or regulatory frameworks can play a critical role in making women's contributions to and experiences with ASM visible. The law can thus amplify women's voices in this space, challenging normative gender roles and improving women's economic opportunities and livelihoods in mining communities (Tobalagba & Vijeyarasa, 2020). Formalizing ASM with a gender perspective creates opportunities to challenge and shift such norms in a sector plagued by gender stereotypes around women's and men's spaces, roles, and behaviors.

There is still much work to be done to understand the effects of gender-responsive legislation on ASM. However, developing reasonable practice regulations on ASM may increase public and private engagement in addressing gender equality. It may help guide future legislation in the extractives industry more broadly (Tobalagba & Vijeyarasa, 2020).

Perhaps the quote below from a woman, Zama Zama, frames it best when it comes to the formalization process and women's involvement. By granting rights, the woman believes that artisanal miners could improve their livelihoods more effectively than through other forms of employment, such as domestic work. This plea underscores the need for a more inclusive approach to mining regulations that considers the socioeconomic realities of artisanal miners and especially women who form part of the sector.

> "We would like for governments to, not necessarily, legalize but to give us some sort of right. Unfortunately now it is hard for us to go and sell it ourselves because of all the challenges we have when we go and sell it ourselves. We get arrested, it gets taken from us. But if those doors could be open for us to sell it ourselves, it would be better. We can make a better living. A better living doing this than in working in people's houses. And it also decreases crime. Because instead of people going stealing in the houses they go and dig for gold."—Esha, Zama Zama

Conclusion

The factors accounting for the limited success of formalization have been widely explained in the literature and include weak institutional capacity, costly and cumbersome bureaucratic processes, and delays that render the entire formalization process unappealing and alienating for many prospective artisanal miners (Banchirigah, 2008; Geenen, 2012; Hilson & Potter, 2005; Maconachie & Hilson, 2011; Tschakert & Singha, 2007).

There is increasing recognition within the Southern African Development Community (SADC) of the need to formalize the ASM sector. In response, the South African government released the draft ASM policy in 2021 and made the policy available for implementation in March 2022. However, the ASM policy can potentially exacerbate policing rather than effectively address and develop the artisanal mining sector. Also, without effectively coupling the formalization process with developmental initiatives, the process will have little to no effect.

For many years, the South African government has largely overlooked the artisanal mining sector, and mining companies have done the same. One could argue that without proactive government leadership, it is unrealistic to expect mining companies to engage with artisanal mining issues. Nevertheless, the importance and role of mining companies relating to the ASM sector cannot be dismissed, as will be discussed in the next chapter. The following chapter examines the dynamics between these two distinct sectors: artisanal miners and their counterparts, LSM companies.

References

Adranyi, E., Stringer, L. C., & Altink, H. (2024). Artisanal and small-scale gold mining governance and cross-sectoral policy coherence in Ghana. *Resources Policy, 96*, 105235. https://doi.org/10.1016/j.resourpol.2024.105235

Adonisi-Kgame, N., Sitefane, M., & Quest, R. (2022). Policy to formalise artisanal and small-scale mining in South Africa published. *Lexology*. https://www.lexology.com/library/detail.aspx?g=0a3d5ffa-3f65-4e4f-ac17-589ee9504811

Afriyie, K., Ganle, J. K., & Adomako, J. A. A. (2016). The good in evil: A discourse analysis of the galamsey industry in Ghana. *Oxford Development Studies, 44*(4), 493–508. https://doi.org/10.1080/13600818.2016.1217984

Banchirigah, S. M. (2008). Challenges with eradicating illegal mining in Ghana: A perspective from the grassroots. *Resources Policy, 33*(1), 29–38. https://doi.org/10.1016/j.resourpol.2007.11.001

Bashwira, M. R., Cuvelier, J., Hilhorst, D., & van der Haar, G. (2014). Not only a man's world: Women's involvement in artisanal mining in eastern DRC. *Resources Policy, 40*, 109–116. https://doi.org/10.1016/j.resourpol.2013.11.002

Bester, V. (2022). Towards a sustainable artisanal gold mining sector in South Africa: Proposed developmental initiatives. *Journal of Rural Studies, 97*, 375–384.

Boafo, G. (2024). *Ghana is Africa's largest gold producer, but it has an illegal mining issue*. Accessed November 2, 2024, from https://theconversation.com/ghana-is-africas-largest-gold-producer-but-it-has-an-illegal-mining-issue-5-essential-reads-238623#:~:text=Ghana%20is%20one%20of%20the,7%25%20of%20gross%20domestic%20product

Boafo, J., Paalo, S. A., & Dotsey, S. (2019). Illicit Chinese small-scale mining in Ghana: Beyond institutional weakness? *Sustainability, 11*(21), 5943. https://doi.org/10.3390/su11215943

Chen, M. (2018). The South African informal sector in international comparative perspective: Theories, data and policies. In F. Fourie (Ed.), *The South African sector: Creating jobs, informal reducing poverty* (1st ed., pp. 26–47). Human Science Research Council.

Clausen, F. (2011). *Property rights theory and the reform of Artisanal and Small-Scale mining in developing countries*. https://ssrn.com/abstract=1907936

Collins, N., & Lawson, L. (2014). *Investigating approaches to working with artisanal and small-scale miners: A compendium of strategies and reports from the field*. Accessed July 5, 2018, from https://www.csrm.uq.edu.au/media/docs/1214/collins-asm-fr-completed-report.pdf

Crawford, G., Agyeyomah, C., Botchwey, G., & Mba, A. (2015). *The impact of Chinese involvement in small-scale gold mining in Ghana*. Research Report. International Growth Centre, LSE. Accessed December 24, 2022, from https://www.theigc.org/wp-content/uploads/2016/08/Crawford-et-al-2015-Final-Report-1.pdf

Davenport, J. (2013). *Digging deep: A history of mining in South Africa*. Jonathan Ball Publishers.

Debrah, A. A., Watson, I., & Quansah, D. P. O. (2014). Comparison between artisanal and small-scale mining in Ghana and South Africa: Lessons learnt and ways forward. *The Journal of the Southern African Institute of Mining and Metallurgy, 114*, 913–921. http://www.scielo.org.za/pdf/jsaimm/v114n11/09.pdf

Department of Mineral Resources, South Africa. (2017). *Small scale mining*. Accessed October 13, 2024, from https://www.dmr.gov.za/mineral-policy-promotion/small-scale-mining

Department of Mineral Resources and Energy. (2021). *Mineral and Petroleum Resources Development Act: Artisanal and small scale-mining pol-*

icy 2022. South African Government. https://www.gov.za/documents/mineral-and-petroleum-resources-development-act-artisanal-and-small-scale-mining-policy

Department of Mineral Resources and Energy (DMRE). (2022). *Publication of the artisanal and small-scale mining policy 2022 for implementation*. Accessed October 13, 2024, from https://www.gov.za/documents/notices/mineral-and-petroleum-resources-development-act-artisanal-and-small-scale-mining

Dzimunya, N., Mapamba, L., Dembetembe, G. G., Dzwiti, K., & Mukono, T. (2018). *Formalization of a roadmap to maximise the contribution of artisanal and small-scale mining in Zimbabwe*. Proceedings of the ASGM Conference, Johannesburg, South Africa, 10–11 September 2018. Accessed March 24, 2023, from https://ir.uz.ac.zw/handle/10646/3833

Fourie, F. (2018). Enabling the forgotten sector: Informal sector realities, policy approaches and formalisation in South Africa. In F. Fourie (Ed.), *The South African sector: Creating jobs, informal reducing poverty* (pp. 439–476). Human Sciences Research Council.

Gauteng City-Region Observatory. (2018). *Mining landscapes of the Gauteng City-region*. Accessed April 4, 2019, from https://www.gcro.ac.za/m/documents/Mining_Landscapes_of_the_GCR_final_web_FA.pdf

Geenen, S. (2012). A dangerous bet: The challenges of formalizing artisanal mining in the Democratic Republic of Congo. *Resources Policy, 37*(3), 322–330. https://doi.org/10.1016/j.resourpol.2012.02.004

Geenen, S., & Claessens, K. (2013). Disputed access to the gold sites in Luhwindja, eastern Democratic Republic of Congo. *Journal of Modern African Studies, 51*(1), 85–108. https://doi.org/10.1017/s0022278x12000559

Gilbert, A. (2002). On the mystery of capital and the myths of Hernando de Soto. What difference does legal title make? *IDPR, 24*(1), 1–19. Accessed March 19, 2022, from https://www.ucl.ac.uk/dpu-projects/drivers_urb_change/urb_infrastructure/pdf_land%20tenure/ESFN_AERUS_Gilbert_mystery_capital.pdf

Hilson, G. (2009). Small-scale mining, poverty and economic development in sub-Saharan Africa: An overview. *Resources Policy, 34*(1–2), 1–5. https://doi.org/10.1016/j.resourpol.2008.12.001

Hilson, G. (2017). Shootings and burning excavators: Some rapid reflections on the Government of Ghana's handling of the informal Galamsey mining 'menace'. *Resources Policy, 54*, 109–116. https://doi.org/10.1016/j.resourpol.2017.09.009

Hilson, G. (2020). 'Formalization bubbles': A blueprint for sustainable artisanal and small-scale mining (ASM) in sub-Saharan Africa. *The Extractive Industries and Society, 7*(4), 1624–1638. https://doi.org/10.1016/j.exis.2020.11.001

Hilson, G., Hilson, A., Maconachie, R., McQuilken, J., & Goumandakoye, H. (2017). Artisanal and small-scale mining (ASM) in sub-Saharan Africa: Re-conceptualizing formalization and 'illegal' activity. *Geoforum, 83*, 80–90. https://doi.org/10.1016/j.geoforum.2017.05.004

Hilson, G., & Potter, C. (2005). Structural adjustment and subsistence industry: Artisanal gold mining in Ghana. *Development and Change, 36*(1), 103–131.

Hilson, G., & Maconachie, R. (2020). Entrepreneurship and innovation in Africa's artisanal and small-scale mining sector: Developments and trajectories. *Journal of Rural Studies, 78*, 149–162.

Hoadley, M., & Limpitlaw, D. (2004). *The artisanal and small scale mining sector & sustainable livelihoods*. Accessed March 20, 2019, from https://www.limpitlawconsulting.com/04Hoadley_Limpitlaw%202004%20The%20artisanal%20and%20small%20scale%20mining%20sector%20&%20sustainable%20livelihoods.pdf

Hruschka, F. (2015). *Comparative analysis of ASM strategies in four countries of Africa, Asia and Latin America (Ghana, Tanzania, Peru and Philippines)*. Accessed June 10, 2019, from http://www.levinsources.com/assets/pages/report-promines-asm-strategies-ghana-tanzania-peru-philippines.pdf

International Conference on the Great Lakes Region (ICGL). (2017). *General overview of the ASM legislations in the Great Lakes Region (Ethiopia and Ghana as benchmark countries)*.

Accessed June 15, 2019, from http://www.icglr-rinr.org/media/attachments/2019/04/23/asm-legal-review-in-the-great-lakes-region_icglr-asm-formalization.pdf

International Crisis Group. (2020). *Mineral concessions: avoiding conflict in DR Congo's Mining Heartland*. Accessed December 24, 2022, from Mineral Concessions: Avoiding Conflict in DR Congo's Mining Heartland | Crisis Group

International Institute for Environment and Development (IIED). (2015). *ASM – LSM – government relations*. Accessed February 22, 2017, from https://iied.org/sites/default/files/pdfs/migrate/G03938.pdf

International Labour Organisation (ILO). (2018). *Women and men in the informal economy: A statistical picture*. Accessed June 28, 2019, from https://www.ilo.org/wcmsp5/groups/public/%2D%2D-dgreports/%2D%2D-dcomm/documents/publication/wcms_626831.pdf

Karaki, K. (2018). *Artisanal gold mining in DRC: time to get down to earth*. Accessed December 24, 2022, from Artisanal gold mining in DRC: Time to get down to earth? - ECDPM

Kingwill, R., Cousins, B., Cousins, T., Hornby, D., Royston, L., & Smit, W. (2006). *Mysteries and myths: De Soto, property and poverty in South Africa*. IIED Gatekeeper Series 124, Institute for Environment and Development. https://www.iied.org/sites/default/files/pdfs/migrate/14517IIED.pdf

Kristof, N. D., & WuDunn, S. (2010). *Half the sky: Turning oppression into opportunity for women worldwide*. Vintage Books.

Kumah, R. (2021). *Ghana needs to rethink its small scale mining strategy. Here's how*. Accessed December 24, 2022, from https://theconversation.com/ghana-needs-to-rethink-its-small-scale-mining-strategy-heres-how-158458

Ledwaba, P. (2017). The status of artisanal and small-scale mining sector in South Africa: Tracking progress. *Journal of the South African Institute of Mining and Metallurgy, 117*(1), 33–40. https://doi.org/10.17159/2411-9717/2017/v117n1a6

Ledwaba, P., & Nhlengetwa, K. (2016). When policy is not enough: Prospects and challenges of artisanal and small-scale mining in South Africa. *Journal of Sustainable Development Law and Policy, 7*(1), 25. https://doi.org/10.4314/jsdlp.v7i1.2

Legal Resources Centre. (2016). *Artisanal Mining Report*. Accessed April 25, 2018, from http://lrc.org.za/wp-content/uploads/pdf/2016_Artisanal%20Mining%20Report_LRC%20Publication.pdf

Levin Sources. (n.d.). *How government engagement and capacity building can help foster the formalisation and development of the ASM sector in Mozambique*. Accessed November 3, 2024, from https://www.levinsources.com/what-we-do/case-studies/government-engagement-capacity-building-formalisation-asm-mozambique

Maconachie, R., & Hilson, G. (2011). Safeguarding livelihoods or exacerbating poverty? Artisanal mining and formalization in West Africa. *Natural Resources Forum, 35*(4), 293–303. https://doi.org/10.1111/j.1477-8947.2011.01407.x

Mcquilken, J., & Hilson, G. (2016). *Artisanal and small-scale gold mining in Ghana: Evidence to inform an 'action dialogue'*. International Institute for Environment and Development. https://www.jstor.org/stable/resrep02683.1

Meredith, M. (2007). *Diamonds, gold and war: The making of South Africa* (1st ed.). Simon & Schuster.

Moraka, N. V., & Van Rensburg, M. J. (2015). Transformation in the South African mining industry – looking beyond the employment equity scorecard. *Journal of the South African Institute of Mining and Metallurgy, 115*(8), 669–678. https://doi.org/10.17159/2411-9717/2015/v115n8a2

Nyamu-Musembi. (2006). *Breathing life into dead theories about property rights: de Soto and land relations in rural Africa - Institute of Development Studies*. Institute of Development Studies. https://www.ids.ac.uk/publications/breathing-life-into-dead-theories-about-property-rights-de-soto-and-land-relations-in-rural-africa/.

Platteau, J.-P. (2015). *Institutions, social norms and economic development*. https://doi.org/10.4324/9780203357606

Robles, M. E., Verbrugge, B., & Geenen, S. (2022). Does formalization make a difference in artisanal and small-scale gold mining (ASGM)? Insights from the Philippines. *The Extractive Industries and Society, 10*, 101078. https://doi.org/10.1016/j.exis.2022.101078

Siegel, S., & Veiga, M. M. (2009). Artisanal and small-scale mining as an extralegal economy: De Soto and the redefinition of "formalization". *Resources Policy, 34*(1–2), 51–56. https://doi.org/10.1016/j.resourpol.2008.02.001

Singo, P., & Seguin, K. (2018). *Best practices: Formalization and due diligence in artisanal and small-scale mining*. Accessed May 20, 2019, from https://www.africaportal.org/documents/18664/IMPACT_ASM-Best-Practices_May-2018-EN-web.pdf

Siwale, A., & Siwale, T. (2017). Has the promise of formalizing artisanal and small-scale mining (ASM) failed? The case of Zambia. *The Extractive Industries and Society, 4*(1), 191–201. https://doi.org/10.1016/j.exis.2016.12.008

South African Human Rights Commission (SAHRC). (2015). *Report of the SAHRC Investigative Hearing Issues and Challenges in Relation to Unregulated Artisanal Underground and Surface Mining Activities in South Africa*. Accessed November 12, 2015, from https://www.sahrc.org.za/home/21/files/Unregulated%20Artisanal%20Underground%20and%20Surface%20Mining%20Activities%20electronic%20version.pdf

Teschner, B. A. (2012). Small-scale mining in Ghana: The government and the Galamsey. *Resources Policy, 37*(3), 308–314. https://doi.org/10.1016/j.resourpol.2012.02.001

The University of British Columbia and Canadian International Resources and Development Institute (CIRDI). (2019). *Artisanal and small-scale mining: Mapping the South African ecosystem*. Accessed November 19, 2021, from https://cirdi.ca/artisanal-and-small-scale-mining-mapping-the-south-african-ecosystem/

Tschakert, P., & Singha, K. (2007). Contaminated identities: Mercury and marginalization in Ghana's artisanal mining sector. *Geoforum, 38*(6), 1304–1321. https://doi.org/10.1016/j.geoforum.2007.05.002

Tobalagba, A., & Vijeyarasa, R. (2020). Engendering regulation of artisanal and small-scale mining: Participation, protection and access to justice. *Third World Quarterly, 41*(10), 1635–1652. https://doi.org/10.1080/01436597.2020.1783995

World Bank. (2020). *2020 State of the artisanal and small-scale mining sector*. Accessed November 19, 2021, from https://documents1.worldbank.org/curated/en/884541630559615834/pdf/Delve-2020-State-of-the-Artisanal-and-Small-Scale-Mining-Sector.pdf

Chapter 8
Artisanal and Small-Scale Mining and Large-Scale Mining Titans

You can't concentrate on doing anything if you are thinking, "What's gonna happen if it doesn't go right?"

—Gladwell (2013)

Introduction

When artisanal and small-scale miners overlap with large-scale mines or vice versa, it is bound to sprout forth a myriad of complex issues. It is never easy for both parties to have constructive conversations with each other. From the artisanal miners' viewpoint, their inferior socioeconomic position makes them skeptical. From the mining companies' standpoint, cynicism about engaging with artisanal miners is understandable, especially in volatile situations. Nonetheless, mining companies have been engaging with the artisanal mining sector in many countries worldwide despite these layers of complexity—some with more success than others.

This chapter explores the tricky and often multifaceted relationship between artisanal mining and large-scale mining (LSM) operations. The dynamic interplay between these two sectors is a subject of great significance within the mining industry discourse.

The chapter examines how LSM companies engage with artisanal miners, shedding light on the diverse strategies employed to navigate this coexistence. These actions hold the potential to catalyze positive transformations. However, such engagements are not without its pitfalls. The chapter explores the potential challenges and risks associated when LSM companies engage with artisanal miners. The chapter concludes by putting forward artisanal and small-scale mining (ASM) intervention areas, followed by discussing two case studies in South Africa and the Democratic Republic of the Congo (DRC) around supporting artisanal mining.

V. Bester, *The Untold Story of Zama Zama Miners in South Africa*, Clinical Sociology: Research and Practice, https://doi.org/10.1007/978-3-031-82783-9_8

A Complex Situationship

Mining companies are hesitant to embrace ASM because of skepticism. The general response of mining companies was, and often still is, to protect their property through private and public security forces (Swiss Agency for Development and Cooperation, 2011). This often leads to police and military raids, arrests, and violent confrontations between security forces and artisanal miners.

In 2002, the International Council on Mining and Metals (ICMM) position was indifferent toward the ASM sector (Swiss Agency for Development and Cooperation, 2011). The ICMM stated that artisanal, small-scale mining and mining legacy issues are important and complex, but they are beyond the capacity of the ICMM. Instead, government and international agencies should lead and address these matters (Swiss Agency for Development and Cooperation, 2011). A few years later, the ICMM changed its position regarding the ASM matter by recognizing the need to engage with the sector proactively. The ICMM stated that the interface between ASM and mining companies should be improved (Swiss Agency for Development and Cooperation, 2011).

In 2009, the World Bank published *Mining Together: Large-Scale Mining Meets Artisanal Mining* (World Bank, 2009). The report was developed to assist the senior executives of LSM companies in managing relationships with individuals who engage in ASM on and around their concessions (Hilson et al., 2020). Published when clashes between these two parties were intensifying across the developing world, *Mining Together* calls on mining corporations to engage, partner, and work with the latter and provides a tentative list of actions on how mining companies should go about doing this (Hilson et al., 2020).

The World Bank is not the only advocate of a collaborative rapport between these two diverse sectors. A more recent report published by the International Institute for Sustainable Development (IGF, 2017), for example, acknowledges that: "*In many parts of the world, ASM and LSM operate in neighboring—and sometimes on the same—concessions*" and stresses that "*their co-existence opens the potential for cooperation.*"

Despite these calls for action, the relations between mining companies and the ASM sector remain complex and primarily conflictive (Bolay, 2014). Conflict over land is common between artisanal miners and mining companies. Artisanal mining regularly takes place on legally owned concessions of mining companies. Mining corporations have claimed their right to mine from the government, but artisanal miners view the situation differently. Sometimes, artisanal miners perceive their claim to work the land as valid based on traditional rights (Geenen & Claessens, 2013). Both sectors comply with different rules, but mining companies have a more advantageous position since they have access to economic resources and are accommodated and favored in legal frameworks (Geenen & Claessens, 2013).

In recent decades, many countries have seen an increase in LSM on the one hand and labor-intensive ASM on the other (Verbrugge, 2015). These two sectors often exist alongside each other, affecting each other; sometimes competing for the same mineral deposits, leading to conflict between the two sectors (Verbrugge, 2017). These hostile interactions have the following two characteristics in common: Firstly, artisanal miners are most often labeled as illegal in these conflictive contexts, and secondly, the core reasons for their labeling as illegal is because of the competition over land tenure and formalization issues (Bolay, 2014). These commonalities contribute to the criminalization of artisanal miners (Bolay, 2014).

This explanation reflects the situation in South Africa. In the South African context, when a relationship emerges between mining companies and artisanal miners, it is tense. This poses a challenge for both mining companies and artisanal miners. The following statements show that, in some cases, no relationship exists, and if one does, it is unhealthy.

> "Our relationship in South Africa with the illegal miners is obviously not a good one, for one simple reason because there is no interaction between the companies and the actual illegal miners."—Bruce, vice president of sustainable development at a mine

Will, a mine manager, had the following to say:

> "Fighting. It's a us vs them. Cause the reality is here on my mine for instance they've disrupt my production so they are preventing us from producing to our full potential. And the things that they do are putting my people in danger. Two reasons why, one: if we catch a guy and he is assisting an illegal they fire him he is fired from the mine, he gets blacklisted and he will never work on a mine again. So his personal career is gone and then they actually in physical danger in terms of water that they put into my tips. I mean the guy at the bottom who needs to operate the box he doesn't know so he opens up and he gets engulfed in mud and water and stuff like that. So they actually, there is a physical danger that they pose as well as personal danger, or career danger."

The last statement indicates that mining companies run a risk when Zama Zama mining occurs on operating mine sites. Managing these risks introduces the idea of risk management.

The International Institute for Environment and Development (IIED) (IIED, 2015) has described how ASM can be dangerous and raise operational risks for a mining company. ASM can be the biggest problem for mining companies by destabilizing the company's social license to operate (Buxton, 2013). Goldfields (2012) stated, for example, that the impacts of the ASM sector threaten to undermine the company's social license to operate.

The notion of a social license to operate developed from within the mining industry as an answer to social risk in the mid-1990s (Boutilier & Thomson, 2011). Nelson (2006) explains that a *"[s]ocial license to operate is a means to earn accountability, credibility, flexibility, and capacity for both stakeholders and industry."* Obtaining a social license to operate means gaining the buy-in and approval of all stakeholders involved who are being affected by a new project (Nelson, 2006).

A mining company's social license comes under threat because of environmental and health problems, conflict with security, allegations of human rights abuse, and the dispute over land and ownership associated with ASM (Buxton, 2013). These impacts were discussed in Chap. 6, and ASM's negative impact on the mining community and companies was pointed out.

The following section examines some motivations and practical actions for mining companies to engage with the artisanal mining sector.

Motivations and Practical Action

The complexity of the nature of the ASM and LSM sectors is apparent. No two situations are ever the same, as several factors shape the scenarios, such as history, conflict, the role of government, and the socioeconomic landscape of the area, to name but a few aspects. Considering all these dynamic parts, the common question is, why and how do mining companies engage with the fluid and complex ASM sector?

Since the artisanal mining sector is associated with social and environmental issues within mining communities, mining companies are increasingly motivated to engage with artisanal miners. The incentive for LSM companies to engage with the ASM sector will differ from operation to operation. Still, it will most likely be classified into at least one of the following broad groups (Business for Social Responsibility, 2014):

1. Risk minimization and security: Where mining companies and ASM activities occur in proximity, ASM can cause health and safety concerns for mining companies. Proactively engaging with miners can mitigate risks.
2. Maximizing development opportunities for communities: Supporting the ASM sector can be essential to a mine's community development strategy. Community development can be described as a mutual process. By helping communities toward sustainability, a company is serving itself, too, because it is creating a healthier and more equal society.
3. Maximization of company benefits: Mining companies can benefit directly when engaging with the ASM sector. For example, when a good relationship is developed between ASM miners and a mining company, mining companies can more easily retain and gain access to exploration licenses and a social license to operate. Additionally, when ASM miners are involved in mine closure planning, the legacy of a mining company and the livelihoods of artisanal miners can be addressed simultaneously. There is room to discover whether artisanal mining can be merged with local development initiatives and community development programs of mining corporations (NeL et al., 2014).

Extracting resources, small or large, involves multifaceted supply and demand networks. While these two sectors might operate at different economies of scale, LSM companies frequently have interests in places where ASM miners are active and vice versa (Kemp & Owen, 2019). Numerous LSM companies have decided to address artisanal mining differently. Ultimately, the nature of the engagements is most likely dependent on the company's vision of sustainable development, capacity, values, and available resources. Also, socio-political factors such as legacy issues around mining practices and land conflict will direct the course of this interface.

Lastly, LSM's role in supporting the ASM sector depends on how the host government views the artisanal miners and how civil society is willing to engage. The roles that LSM can uptake are highly dependent and will vary based on the local context (World Gold Council, 2022). Still, as part of their role in supporting dialogue and engagement, LSM companies may consider several options to support artisanal mining, which can include the following (World Gold Council, 2022):

- A mining company can advocate for formalizing and legitimizing ASM activities. Also, mining companies can support governments and other actors in establishing legal and regulatory frameworks for the ASM sector.
- Mining companies can work with governments and other stakeholders to combat breaches of human rights associated with abusive ASM activities. Furthermore, a mining company can promote awareness of and access to technologies that reduce environmental and worker safety risks. Also, an LSM company can promote awareness of the dangers of the worst forms of child labor, modern slavery, and gender-based violence, all issues in the artisanal mining sector.
- Mining companies can recognize artisanal and small-scale miners as stakeholders, include them in site-level impact assessments and community engagement practices, and consider them part of social investment priorities.
- Another way for LSM companies to support the ASM sector is to compensate established and small-scale miners reasonably. If a large-scale mine legitimately physically or economically displaces ASM miners, alternative livelihood initiatives should be supported.
- Another positive interactive activity is for a mining company to implement the UN Guiding Principles on Business and Human Rights in their dealings with ASM and local communities. This may also include the provision of appropriate grievance mechanisms.
- A mining company can also consider facilitating access to legal markets for legitimate ASM production, including where environmental, social, and governance due diligence requirements have been met, and by accepting gold-bearing material for processing.
- LSM companies can also pass information to national authorities and national and international law enforcement agencies about the doings of criminal organizations intended to intimidate, extort, or exploit ASM groups.
- LSM companies can play an essential role in reviewing the potential for assisting with mercury replacement and the facilitation of capacity-building to improve ASM practices.
- Lastly, an LSM company can also consider concession areas not viable for LSM activities. This is, of course, dependent on the government's stance toward artisanal mining. A mining company can also seek to work with local authorities to pre-empt any related influx of miners that might undermine the sustainability of such schemes and the social fabric of the local area.

In September 2024, the World Bank unveiled an ambitious new vision to unlock the potential of ASM. The report identified nine key intervention areas based on case studies, interviews, and a literature review (World Bank, 2024). The rationale behind the World Bank's ASM framework is part of a broader reflective process to evaluate past interventions and approaches, adapt to emerging global development agendas, and to enhance support for their client countries.

Practically, the framework examines and offers various ASM intervention strategies that can be implemented to address ASM more effectively, as identified by the World Bank. The areas of intervention include the following broad thematic themes: *Ensure responsive and appropriate regulatory frameworks*; *Expand the application of progressive mine site standards*; *Enable gender equality across the value chain*; *Support appropriate decentralization and local governance models*; *Advance environmental stewardship and security*; *Facilitate financial inclusion*; *Support infrastructure to access legal markets*; *Build the organizational capacity of ASM miners; and Deliver extension services for miners.*

Each of the nine areas is accompanied by practical actions that can be followed. The World Bank's (2024) ASM framework concludes by emphasizing an alternative approach to support ASM by integrating regulated financing, boosting market access, and promoting capacity building. Achieving sustainable ASM requires scaling through partnerships with public and private sectors. Safety, well-being, environmental stewardship, and increased earnings are key success metrics (World Bank, 2024).

Nonetheless, it is not always easy for a mining company to interact with the ASM sector, as interactions are not solely dependent on its willingness or resources. Several complications have been documented in the literature when mining companies incorporate artisanal mining into their local development strategies. The following section provides an overview of such.

Downsides to Be Aware of When Engaging with the Artisanal Mining Sector

Most often, alternative livelihood strategies or projects are proposed to address the presence of artisanal miners at a given site. However, alternative livelihood projects fail because artisanal miners find the proposed livelihood initiative unfitting. This mismatch is often the result of no dialogue between mining companies and the ASM sector. The unsuitability of many alternatives, such as promoting agricultural jobs where they are unfit for the context, prevents artisanal miners from abandoning their activities (Banchirigah, 2008).

Artisanal miners will not readily leave their livelihood without any viable, profitable alternative they feel comfortable with. Also, those initiatives that involve pure force to clamp down on artisanal mining activities are unsustainable and ineffective.

Kemp and Owen (2019) identified the following dilemmas and issues that LSM companies might encounter when interacting, or when a mining company is willing to interact, with the ASM sector:

- Initiating engagement: Approaching ASM as a sector and considering the roles, responsibilities, and networks of different groups within that sector makes sense as a principle. A detailed perspective of the specific operating context should be produced at the earliest feasible stage of a project's lifespan. It is suggested that LSM companies in some regions must first engage with the government to avoid being accused of facilitating illicit behavior.
- Land, boundaries, and encroachment: Land encroachment is likely the most contentious issue between the two sectors. Understanding the official and informal processes associated with various types of land tenure is essential for resolving challenges. The structure of land relations reveals how distinct sets of official and informal rights are allocated, to whom they are allocated, and how land rights and entitlements are governed through diverse governance systems. It also applies to ASM in many circumstances, where the right to access and land use is subject to traditional authority clearance.
- Reduced project value: The interface between LSM and ASM can cause problems, including sterilizing the resource. LSM companies may incur enormous security costs and hazards when owning exclusive leasing zones. In the early stages of project development, high-security expenses lower the return on investment and raise operating costs later. Security interventions can heighten the risk of human rights violations by the state or abuses by private security forces or company security personnel. Similarly, law and order difficulties in and near settlements can jeopardize employees' and their families' safety and security.
- Compensating for loss and disruption: Cash compensation should be carefully considered for dealing with the effects of displacement and loss of livelihood. While compensation sums should reflect the worth of the resources and possibilities lost, some large-scale operators will try to avoid or minimize compensation payments, especially if the action is not sanctioned by the government and is considered illegal, as is often the case in artisanal mining.
- Valuing and ASM resources: The difficulty of assessing the worth of artisanal-mined resources hinders equitable pay. Parties must arrive at an estimated value of the geological resource if LSM companies seek to compensate for the loss. Without a model to guide the parties through the valuation process, the problem of one party saying the resource should be valued at a much higher rate while the other side says the opposite will always exist.
- Managing livelihood dimensions: The livelihood features of LSM and induced displacement have been noted as a challenging practice area in the scholarly literature. The lack of credible valuation methodologies to determine compensation rates and the lack of formal recognition in the ASM sector add to the complexity. These unresolved concerns have real-world implications for all parties involved. The notion that displaced ASM miners will return to subsistence agriculture is widely used to develop livelihood restoration projects. There is a

lack of understanding among LSM companies that ASM is, in most cases, already an alternative livelihood. Alternative livelihood programs, by and large, are ineffectual, not only because displaced people find the replacement livelihood unappealing but also because there is a lack of consensus on what people are entitled to if the resource is lost to LSM. A fundamental difference exists between valuing an underlying asset, such as an alluvial deposit, and an overlying asset, such as agriculture.

- Responding to in-migration and speculation: In-migration is often linked to ASM and is portrayed as dangerous to industrial mining ventures. Similarly, in-migration, when LSM and ASM coexist, might be more difficult to avoid or manage once individuals start moving and settling in. The timing, intensity, and patterns of in-migration influence the spectrum of reactions available in any given environment.
- Mapping illegal activities: For LSM companies, stakeholder mapping is a significant difficulty in illegal or criminal activity. The mapping of legal, informal, illicit, and criminal groups and elements is required to comprehend the ASM sector. Without correctly mapping these activities, LSM companies may unintentionally help criminal groups or import criminal elements into their supply chain without a clear line of sight.
- Understanding economic networks: Local people may have entered the LSM supply chain through employment or procurement after being evicted from their land. Or, residents may exit the LSM supply chain and re-engage in ASM during a downturn or when LSM companies near closure. For example, some ASM miners in Tanzania participate in both sectors' supply chains at various times, depending on assessed security and livelihood concerns (Jønsson & Bryceson, 2009). Understanding these movement patterns is essential for identifying the push and pull factors that motivate artisanal miners into or out of ASM.

Besides these difficulties and challenges, one reason programs targeting artisanal mining often fail is that mining companies frequently develop community initiatives without involving the communities they aim to support. Several corporate-sponsored programs target communities near mining companies, and not artisanal miners (Banchirigah, 2008). This mis-targeting assumes that the mining community is homogeneous and bounded (Bolay, 2014). Numerous scholars alluded that when addressing the challenging ASM sector, the complex heterogenous nature of the ASM sector must be recognized (Banchirigah, 2008; Verbrugge, 2016; Verbrugge & Besmanos, 2016). Most often, initiatives from mining companies and governments fail because of an erroneous assumption that the sector is homogeneous (Banchirigah, 2008; Verbrugge & Besmanos, 2016). A lack of baseline research fuels beliefs about the homogeneity of artisanal miners. In many instances failed ASM interventions could have been avoided if a proper baseline study about the artisanal miners had been conducted (Hilson, 2007).

A classic example of an initiative that failed because of the lack of a proper baseline study is the Shamva Mining Center in Zimbabwe. Established in 1989, the Shamva Mining Center in Zimbabwe began as a pilot project to offer technical and financial support to ASM miners in the region. The Shamva Mining Center was initially built to serve 40 local gold miners to offer as a centralized processing point. But, shortly after its construction, it started to process ore from over 500 artisanal and small-scale operations. The capacity of Shamva Mining Center could not meet the growing needs of the miners. Ultimately, the miners had to wait 3–6 weeks for their minerals to be treated.

It is argued that this problem could have been avoided if a baseline study had been conducted in ASM communities. Focusing on the size of the artisanal mining populations, their different skill levels, and their various demands for services would have put the developers and designers of the central processing center in a better position to meet the local needs and demands. This poor understanding of the artisanal mining community results in inappropriate support initiatives. The problem with the homogeneous assumption, aggravated by having no baseline research, is that corporate social strategies or community development projects from mining companies add to social fragmentation at a local level, leaving no room to consider the local group dynamics (Bolay, 2014).

Collaboration initiatives are sometimes another proposed initiative to support the ASM sector. Some interviewees, during the discussions I had, also alluded to collaboration between artisanal miners and mining companies to curb exploitation and extortion; however, Hilson et al. (2020) caution against considering cohabitation ideas between artisanal miners and mining companies.

Cohabitation partnerships are frequently optimal under specific circumstances and are only sustainable for a limited time. According to Hilson et al. (2020), cohabitation is an unsustainable strategy because of fluctuations in the gold price and the volatility of mine ownership.

Instead, a self-sufficient coexistence should be sought. Because of the significant differences between the two sectors, separate policy treatment is required to allow each to thrive independently (Hilson et al., 2020).

However, the absence of an international framework for coexistence between the two industries raises substantial issues and has resulted in some of the most severe human rights breaches, as seen in the Congolese mining industry (Deberdt, 2021). In many instances, the convergence between the LSM practices, governmental action, a country's legal system, and the array of challenges in the ASM sector make any coexistence between the two sectors challenging. Nevertheless, a solution is attainable if all the stakeholders are prepared to work together and adopt a strategy that will benefit all parties, including the artisanal miners (Deberdt, 2021).

Involving civil society is one way to balance this multi-stakeholder relationship. According to Echavarria (2014), a third party that is relatively neutral and promotes transparency, such as academia or a non-governmental organization, can help navigate this space. According to research, there is a significant power imbalance between mining companies and artisanal miners. When a mining company wants to engage with the ASM sector, a mediator's role is critical to leveling the playing field (Andrew & Hilson, 2003).

Case Studies

In South Africa, efforts to address artisanal miners developmentally have been minimal because of the lack of legal status for Zama Zamas. However, in June 2018, the then Department of Mineral Resources (DMR) granted mining permits to Zama Zamas, in Kimberley, Northern Cape. This initiative, involving the Kimberley Ekapa Mining Joint Venture (JV), marked a historic step toward integrating artisanal miners into the legal mining framework and providing support. The miners had access to 500 ha of land to mine legally, which was expected to improve their livelihoods and reduce illegal mining activities (Khumalo, 2018).

Unfortunately, the project faced significant challenges. In November 2019, violence erupted, involving artisanal miners excluded from the initiative. These miners attacked the infrastructure and those participating in the project (Heiberg & Reid, 2019).

It is important to note that the model or agreement between the Zama Zamas, Kimberley Ekapa Mining JV, and the then-DMR is unlikely to be replicable due to the diverse nature of Zama Zamas, including the types of ore they mine. Nevertheless, this initiative and its complications can still provide valuable insights for developing similar projects with multi-stakeholder partnerships.

Building on this, Petra Diamonds, part of the Kimberley Ekapa Mining JV, expanded its artisanal mining scope in October 2019 to its Koffiefontein operation in the Free State Province. At that time, the company aimed to implement the lessons learned from the Kimberley initiative to assist artisanal miners in the new area (Jamasmie, 2019).

Turning now to the second case study. An important emerging aspect of LSM and ASM engagement is the traceability of the minerals. In recent years, the debate around responsible resourcing has boomed. There is now more than ever a massive focus on supply chains and human rights violations within mineral supply chains; for that reason, the next initiative that will be discussed is about responsible sourcing, which was put in place in the DRC. It has been reported that in the DRC, ASM with active and responsible resourcing programs is increasingly becoming a focus of research (Deberdt, 2021).

For that reason, traceability schemes have become an essential component of multi-stakeholder responses to various problems in the mining sector, particularly in the ASM sector, where exploitative practices frequently result in human rights violations and prevent miners from exiting poverty cycles (Sovacool, 2021). These schemes may, however, have some negative consequences: for example, Deberdt (2021) contends that the due diligence and traceability approach *"reinforces an already existing trend of militarization (or para-militarization) of mine sites, which led to significant clashes between security forces and artisanal miners"* in the DRC's cobalt sector.

The Mutoshi Pilot Project (in the DRC) may serve as a model for imagining future responsible resourcing projects. The Mutoshi Pilot Project was established in January 2018 by multiple stakeholders. Chemaf, Trafigura Group, Pact, Kumi

Consulting, and COMIAKOL collaborated to implement the project, which was situated outside of Kolwezi in the Lualaba Province. The project aimed to build the capacity of on-site suppliers, workers, and stakeholders to maintain Trafigura's Expectations and other international standards, as well as improve socioeconomic conditions for the artisanal miners and their families (Johansson de Silva et al., 2019).

The COMIAKOL cooperative was chosen as the artisanal mining partner since it drew its members from the local community and was already in the process of obtaining a mining license. When the project started, about 5000 artisanal miners registered as members of the COMIAKOL cooperative. For security reasons, no more than 5000 artisanal miners were allowed on-site on any given day. However, the number of workers on-site varied, as most workers operate on a bi-weekly shift (Johansson de Silva et al., 2019). To enter the Mutoshi site, all artisanal miners had to be members of COMIAKOL and carry with them identity cards explicitly developed for the site. The site was also fenced off to prevent external miners or family members from accessing the mine (Johansson de Silva et al., 2019).

Working with Trafigura, Pact has overseen designing the scope of the pilot project and provided technical support and on-the-job training on occupational health and safety, roles and responsibilities, specialized cooperative training, security, and human rights, ASM labor transitioning and resilience, as well as data collection and monitoring of the project. Assessments conducted by Pact were reinforced quarterly by Trafigura's appointment of the responsible supply chain consultancy, Kumi, whose mandate was to scrutinize the project in-depth and report back on performance to Trafigura, Chemaf, and Pact, as well as selected downstream partners.

Kumi assessed the project's performance against a set of standards, including the OECD Due Diligence Guidance for Responsible Supply Chains of Minerals from Conflict-Affected and High-Risk Areas; the International Finance Corporation's (IFC's) Performance Standards; Trafigura's Business Principles for Health, Safety, Environment, and Community Performance; and Trafigura's Responsible Sourcing and Supply Chain Expectations, plus the expectations designed explicitly for this semi-mechanized context (Johansson de Silva et al., 2019).

Chemaf and COMIAKOL have negotiated a price agreement for the cobalt produced at the Mutoshi mine, and Chemaf committed to purchasing all the cobalt produced by the COMIAKOL members.

In addition, Chemaf oversaw safety procedures, health services, safety equipment, community liaisons, and managing the excavators. In this context, excavators were required to remove waste and ensure that at no point did pits dug by artisanal miners exceed 10 m in depth. DRC's State Regulator for ASM Service for Assistance and Supervision of ASM (SAEMAPE) conducted safety inspections. The various stakeholders monitored different aspects of safety, including child labor (Johansson de Silva et al., 2019).

The project positively impacted the artisanal miners by allowing the miners to contribute significantly to the local economy. The key positive outcomes of the project can be listed as follows (Johansson de Silva et al., 2019):

- The project had significant positive impacts on the productivity of the artisanal miners when they were on-site.
- Overburden stripping, provision of geological information, and better workplace organization helped increase the outputs. However, several services provided by the project were withdrawn during the first half of 2019, which caused a severe relapse of productivity for the participating artisanal miners.
- The project has also helped reduce the artisanal miners' operating costs, for example, transporting produce to trading places and eliminating bribes paid to corrupt military and security forces.
- Working conditions of the participating artisanal miners improved remarkably, for example, dangerous situations were made safer, and the artisanal miners worked fewer hours.
- The artisanal miners felt safer at their workplaces and could spend more time with their families. These improvements have been sustained during price fluctuations.
- The Mutoshi mine became a safer, cleaner environment, and accessible health-care services significantly improved the health of the participating artisanal miners. The artisanal miners were less likely to lose working days because of illness or injury.
- The project had a noticeable impact on the local economy, including creating new businesses in response to higher demand for goods and services: Miners spend an estimated three-quarters of their total spending on consumption items on goods that have been locally produced.
- Female artisanal miners participating in the project had a greater diversity of roles and earned about two and a half times more than their counterparts working in mines outside the project site. Following the project launch, some women moved from female-dominated occupations (washing) into typically male-dominated occupations (operating diggers, etc.) that offer higher earnings. They also reported that workplace harassment from male colleagues decreased and that improved site sanitation (clean water and toilets) has positively impacted their health and well-being.
- Lastly, with the departure of military forces that were previously deployed to protect the site from artisanal miners, many new businesses started in the area and were owned by women, reflecting a more secure working environment and more significant opportunity for female business owners.

However, the project's implementation and impact were severely affected by a dip in the world market price of cobalt. In response to a severe drop in turnover and lack of liquidity, Chemaf withdrew specific project deliverables, such as regular and frequent payments to the miners, stripping off the mine site, and replacing worn or lost safety equipment. Suspended in March 2020 because of the COVID-19 pandemic, the project ended in December 2020 because of a combination of factors, including the impacts of the pandemic, the redirection of priorities, and tensions between partners (Deberdt, 2021).

Despite the project's termination, it offers important lessons and illustrates that small-scale and artisanal mining could be commercially viable while addressing extensive material and reputational hazards (Deberdt, 2021). The project is also a prime example of establishing a solid foundation for collaboration between ASM and LSM, leading to significant benefits for the miners involved (Deberdt, 2021).

However, the costs remain an essential barrier to scaling up (Deberdt, 2021). Also, the level of informality, the attitude of the military and police, including the Police des Mines et des Hydrocarbures (PMH) (mining police), and the agreements between artisanal miners and private security firms, significantly complicated the access to the sites (Deberdt, 2021). Despite these challenges, the Mutoshi Pilot Project remains a prime example of the formation of grassroots-based cooperatives to produce conflict-free minerals.

Conclusion

There exist various reasons why mining companies should engage with the artisanal mining sector. One key driver is the realization that ASM threatens a mining operation's social license, prompting mining companies to consider how best to address this sector. Many mining companies worldwide recognize the potential benefits of adopting a collaborative approach to tackle ASM (Verbrugge, 2017). While some mining firms have begun to view ASM as a social issue and have engaged with and supported its development, there remains a notable gap in what mining companies can do to promote the ASM sector.

This oversight is particularly evident in South Africa, where community development initiatives directed at addressing ASM are limited, if they exist at all.

In conclusion, the relationship between artisanal and small-scale miners and LSM operations is complex and multifaceted, presenting both opportunities and challenges. While collaboration between these sectors has seen varied levels of success worldwide, addressing the socioeconomic disparities and potential conflicts remains crucial. The proposed interventional areas highlighted by the World Bank (2024) and the two case studies presented in this chapter showcase promising approaches to support artisanal mining. However, caution must be exercised to navigate the inherent risks and challenges. By fostering a careful, balanced engagement, there is potential to drive positive transformation and to create a more inclusive, sustainable mining industry.

The next chapter, which serves as the conclusion of this book, outlines proposed initiatives to address artisanal mining in South Africa.

References

Andrew, J. S., & Hilson, G. M. (2003). Land use disputes between small- and large-scale miners: improving conflict management. In G. M. Hilson (Ed.), *The socio-economic impacts of artisanal and small-scale mining in developing countries* (pp. 25–42). Balkema Publishers.

Banchirigah, S. M. (2008). Challenges with eradicating illegal mining in Ghana: A perspective from the grassroots. *Resources Policy, 33*(1), 29–38. https://doi.org/10.1016/j.resourpol.2007.11.001

Bolay, M. (2014). When miners become "foreigners": Competing categorizations within gold mining spaces in Guinea. *Resources Policy, 40*, 117–127. https://doi.org/10.1016/j.resourpol.2014.02.001

Boutilier, R., & Thomson, I. (2011). *Modelling and measuring the social license to operate: Fruits of a dialogue between theory and practice.* International Mine Management. Accessed March 16, 2020, from https://socialicense.com/publications/Modelling%20and%20Measuring%20the%20SLO.pdf

Business for Social Responsibility (BSR). (2014). *How can business contribute to the ethical mining of conflict minerals.* Accessed December 20, 2018, from https://www.bsr.org/reports/bsr_Ethical_Mining_Conflict_Minerals.pdf

Buxton, A. (2013). *Responding to the challenge of artisanal and small-scale mining. How can knowledge networks help?* Accessed August 15, 2015, from http://pubs.iied.org/16532IIED.html

Deberdt, R. (2021). The Democratic Republic of the Congo (DRC)'s response to artisanal cobalt mining: The Entreprise Générale du Cobalt (EGC). *The Extractive Industries and Society, 8*(4), 101013. https://doi.org/10.1016/j.exis.2021.101013

Echavarria, C. (2014). *'What is legal'? Formalising artisanal and small-scale mining in Colombia.* Accessed September 9, 2020, from https://pubs.iied.org/pdfs/16565IIED.pdf

Geenen, S., & Claessens, K. (2013). Disputed access to the gold sites in Luhwindja, eastern Democratic Republic of Congo. *Journal of Modern African Studies, 51*(1), 85–108. https://doi.org/10.1017/s0022278x12000559

Gladwell, M. (2013). *David and Goliath: Underdogs, misfits, and the art of battling giants.* Allen Lane.

Goldfields. (2012). *Golden age.* Accessed January 7, 2019, from http://www.goldfields.co.za/pdf/media/publications/group/2012/april-golden-age-news.pdf

Heiberg, T., & Reid, H. (2019, November 11). In Kimberley, the world's diamond capital, illicit mining fight flounders. *Reuters.* Accessed March 20, 2020, from https://www.reuters.com/article/us-safrica-mining-illegal/in-kimberley-the-worlds-diamond-capital-illicit-mining-fight-flounders-idUSKBN1XL0J0

Hilson, G. (2007). What is wrong with the Global Support Facility for small-scale mining? *Progress in Development Studies, 7*(3), 235–249. https://doi.org/10.1177/146499340700700304

Hilson, G., Sauerwein, T., & Owen, J. R. (2020). Large and artisanal scale mine development: The case for autonomous co-existence. *World Development, 130*, 104919. https://doi.org/10.1016/j.worlddev.2020.104919

Intergovernmental Forum (IGF). (2017). *IGF guidance for governments. Managing artisanal and small-scale mining.* Accessed February 10, 2017, from http://www.iisd.org/sites/default/files/publications/igf-guidance-for-governments-asm_0.pdf

International Institute for Environment and Development (IIED). (2015). *ASM – LSM – Government relations.* Accessed February 22, 2017, from https://iied.org/sites/default/files/pdfs/migrate/G03938.pdf

Jamasmie, C. (2019, October 31). Petra Diamonds opens tailings to artisanal miners in South Africa. MINING.COM. Accessed December 20, 2019, from https://www.mining.com/petra-diamonds-opens-tailings-to-artisanal-miners-in-south-africa/

Johansson de Silva, S., Strauss, T., & Morisho, N. (2019). *The Mutoshi Pilot project.* Trafigura. https://www.trafigura.com/media/1qnci1so/2019_trafigura_the_mutoshi-pilot_project.pdf

Jønsson, J. B., & Bryceson, D. F. (2009). Rushing for gold: Mobility and Small-Scale mining in East Africa. *Development and Change, 40*(2), 249–279. https://doi.org/10.1111/j.1467-7660.2009.01514.x

Khumalo, S. (2018, June 7). Thousands of zama-zamas receive mining permits. Mail & Guardian. Accessed December 20, 2019, from https://mg.co.za/article/2018-06-07-thousands-of-zama-zamas-receive-mining-permits/

Kemp, D., & Owen, J. R. (2019). Characterising the interface between large and small-scale mining. *The Extractive Industries and Society, 6*(4), 1091–1100. https://doi.org/10.1016/j.exis.2019.07.002

NeL, E., Binns, T., & Gibb, M. (2014). Community development at the coal face: Networks and sustainability among artisanal mining communities in Indwe, Eastern Cape Province, South Africa. *The Geographical Journal, 180*(2), 175–184. http://www.jstor.org/stable/43868602

Nelson, J. L. (2006). Social license to operate. *International Journal of Mining, Reclamation and Environment, 20*(3), 161–162.

Sovacool, B. K. (2021). When subterranean slavery supports sustainability transitions? Power, patriarchy, and child labor in artisanal Congolese cobalt mining. *The Extractive Industries and Society, 8*(1), 271–293.

Swiss Agency for Development and Cooperation. (2011). *SDC experiences with formalisation and responsible environmental practices in artisanal and small-scale gold mining in Latin America and Asia (Mongolia)*. Accessed July 20, 2019, from https://www.unep.org/globalmercurypartnership/resources/report/sdc-experiences-formalization-and-responsible-environmental-practices-artisanal

Verbrugge, B. (2015). The economic logic of persistent informality: Artisanal and Small-Scale mining in the Southern Philippines. *Development and Change, 46*(5), 1023–1046. https://doi.org/10.1111/dech.12189

Verbrugge, B. (2016). Voices from below: Artisanal- and small-scale mining as a product and catalyst of rural transformation. *Journal of Rural Studies, 47*, 108–116. https://doi.org/10.1016/j.jrurstud.2016.07.025

Verbrugge, B. (2017). Towards a negotiated solution to conflicts between large-scale and small-scale miners? The Acupan contract mining project in the Philippines. *The Extractive Industries and Society, 4*(2), 352–360. https://doi.org/10.1016/j.exis.2016.10.011

Verbrugge, B., & Besmanos, B. (2016). Formalizing artisanal and small-scale mining: Whither the workforce? *Resources Policy, 47*, 134–141.

World Bank. (2009). *Mining together large-scale mining meets artisanal mining: A guide for action*. Accessed December 24, 2022, from file:///C:/Users/videt/Downloads/686190ESW0P1120ng0Together0HD0final%20(4).pdf

World Bank. (2024). *Achieving sustainable and inclusive artisanal and small-scale mining (ASM): A renewed framework for World Bank engagement*. Accessed November 3, 2024, from https://openknowledge.worldbank.org/server/api/core/bitstreams/cab95ce4-781f-42df-9c4d-912c55834c1b/content

World Gold Council. (2022). *Lessons learned on managing the interface between large-scale and artisanal and small-scale gold mining*. Accessed December 24, 2022, from https://www.gold.org/esg/artisanal-and-small-scale-gold-mining#registration-type=google&just-verified=1

Chapter 9
At a Crossroads

> *Do not look the other way; do not hesitate. Recognize that the world is hungry for action, not words. Act with courage and vision.*
>
> —Nelson Mandela (as cited by the United Nations, 2023)

Introduction

A country's political, socioeconomic, and historical aspects are integral to the broader context in which artisanal miners operate. These factors cannot be overlooked.

In South Africa, mining companies have had a long-standing reputation for inflicting enormous social harm on the surrounding communities and the miners from whom they extracted their labor. A comprehensive understanding of the history of South African mining makes it easier to see how the South African government and mining companies from the pre-apartheid and apartheid era have excluded specific groups from accessing mineral rights.

The actions of the early 1900s speak of atrocities, bearing in mind the migrant labor system, the mining compounds and hostels, and the various Acts that aimed to foster racial exclusion. It would be limiting to consider Zama Zama mining without understanding and recognizing the deep racial and socioeconomic divide that the mining industry co-created with the pre-apartheid and apartheid governments. Such an understanding will also shed light on power dynamics and who gets to have, and always had, a more significant say in narratives and policy making.

Acknowledging the apartheid-era's actions and social injustices, the Mineral and Petroleum Resources Development Act (MPRDA) of 2002 was implemented to address the historical harms and promote the overall artisanal and small-scale mining (ASM) sector in South Africa. One would have hoped that the post-apartheid era would have brought a new dawn for many, rectifying the injustices from the past. Regrettably, as can be seen from the neglect of addressing the country's artisanal mining sector, the matter of economic inclusion and addressing the wrongs from the past is once again being ignored or even met with force.

This brings us to the crucial point of addressing the historical and social injustices and re-evaluating Zama Zama mining within a developmental framework. This

V. Bester, *The Untold Story of Zama Zama Miners in South Africa*, Clinical Sociology: Research and Practice, https://doi.org/10.1007/978-3-031-82783-9_9

shift is long overdue. As the concluding chapter of the book, this chapter outlines initiatives that could be implemented to address Zama Zama mining in a more sustainable manner. Hopefully, these initiatives will help facilitate the redistribution of mineral resources and enhance access for individuals who have been socioeconomically marginalized up to this point.

A Shift to Redistribution of Resources

With numerous mining communities relying on large-scale mining (LSM) economies in areas, it is critical to investigate artisanal mining as an alternative livelihood strategy. The Zama Zama route, for many, became a means of earning a living because of a lack of job opportunities in the formal sector (Action Aid, 2019; Bester, 2022). However, the South African government and industry players frequently overlook this sector by focusing on the adverse effects of Zama Zama mining, rather than addressing the challenges to improve the sector's opportunities for sustainable development.

Yet, numerous initiatives worldwide have been developed to address and support the socioeconomic potential of this sector (IGF, 2017; Zvarivadza, 2018). These initiatives promote local development and acknowledge mining companies' role in developing the artisanal mining sector (Calderon et al., 2016; IGF, 2017). Due to the Zama Zama's lack of legal status, little effort has been made in South Africa to address artisanal miners developmentally.

Exploring practical ways to address the artisanal mining sector is more vital than ever, given the South African ASM policy, which has been up for implementation since March 2022. As highlighted in Chap. 7, the policy is fraught with pitfalls and often lacks clarity, particularly regarding the role of women in the artisanal mining sector. Another risk is that there might exist too much reliance on the formalization process to address the current Zama Zama issue. Formalization efforts have mostly been unsuccessful and are a challenging process. Thus, exploring alternative initiatives to assist the sector and enhance the formalization process is essential.

Many scholars have pointed out that mere legal changes will add little value to developing a more sustainable artisanal mining sector. When exploring ways to develop and support the sector, a diverse set of supporting initiatives and multiple stakeholder partnerships are required. Mining company representatives, government officials, civil society members, artisanal miners, and international development agencies should all be stakeholders (Aubynn, 2009).

The following sections provide initiatives, from a multi-stakeholder partnership perspective, to address artisanal mining in the South African context. This section draws on the findings of a research paper I published in 2023 (Bester, 2023). The suggested initiatives can also contribute to the formalization process of artisanal mining in South Africa, positioning the process as a developmental pathway.

Changing Mindsets

During the interviews I conducted, many interviewees stated that to address artisanal mining, mining companies, governments, and regulatory bodies such as the Minerals Council should change their mindset. Participants suggested that rather than viewing the Zama Zamas as criminals, a shift in perspective is needed. Development initiatives will begin to emerge because of a different perspective on the sector, as one interviewee puts it:

> "I think the one big problem, or the starting point in my view, should be to deal with the artisanal miners and illegal miners in a developmental way. That change of thinking is required. At the moment it's confrontational. It's criminalizing them".—Jack, key informant

A civil society worker commented that a mind shift could allow for the redistribution of resources that will ultimately close the poverty gap:

> "And you know, again, when you talk about the mind shift and you talk about redistribution of resources, you talk about closing the poverty gap".—Nick, key informant

Given the preceding statements, a shift in mindset will allow mining companies and government entities to view and treat artisanal mining with less force. Nyoni (2017) identifies the same need in his research. According to Nyoni (2017), research participants' narratives indicate that some Zama Zamas are former gang members who decided it was better to abandon their criminal activities and pursue artisanal gold mining, which they regard as a *"clean"* source of income. As a result, changing perceptions about artisanal mining activities is critical. This is especially important in the government's approach, which should regard this sector as a legitimate source of income whose regulation can benefit all actors (Nyoni, 2017).

It would be naïve to expect mining companies to recognize the value of Zama Zama mining suddenly. Understanding the sector will take time and may not occur until mining companies anticipate an incentive. Reduced risk from artisanal mining for the company and mining communities can encourage companies to address Zama Zama mining sustainably. However, this mindset shift necessitates a process of self-evaluation and a mature assessment of the effects of mining operations. According to the International Institute for Environment and Development (IIED), one of the challenges for businesses is determining whether ASM should be viewed as a security or social issue (IIED, 2015).

The IIED (2015) explains that ASM could be a security threat and operational risk. For mining companies, to treat artisanal mining as a social issue requires a shift in thinking about how to handle the artisanal miners differently. Also, a different narrative will only develop when another type of thinking is created around artisanal mining in the formal mining space, and each party can move away from seeing the other as a threat.

Training in Health and Safety Issues

When artisanal mining occurs on concessions where a company still holds a mining right, Zama Zama miners and formal mine workers face several health and safety risks, which pose risks for mining companies. According to Martin (2019), the Zama Zama situation is as follows: Illegal artisanal gold mining occurs underground, primarily in derelict and abandoned mines and operating industrial shafts, in exploitative and dangerous conditions, almost entirely under the control and direction of sophisticated criminal syndicates. While desperation and difficult socioeconomic conditions frequently motivated the Zama Zamas, the syndicates' motivations are more dishonest and malevolent. Service providers and civil society view health and safety as the industry's greatest concerns. Working together with artisanal miners, mining companies could address this problem.

During the discussions, it was suggested, more than once, that health and safety campaigns in areas where Zama Zama activity is occurring could reduce risks and challenges for mining companies and miners. During a conversation with a Zama Zama miner, he indicated that they are themselves unaware of the dangers of mining:

> "People are dying. Why are people dying? People they don't have knowledge they don't have, they don't know actually where it's dangerous. Where there is [a] barricade, where there is no entry, no go area. We are using these places that's closed for a long time you see".—Tim, Zama Zama

The above quotation underscores the need for capacity building among the miners. Initiatives that address health and safety will empower the miners. During two separate conversations with a civil society member and a Sustainable Development manager from a mining company, both mentioned solution-based initiatives related to health and safety and said the following:

> "That is why you also need health officers from the government to actually to check up on the health of these guys".—Danny, civil society

> "To eliminate mercury and cyanide use totally. And there are technologies, gravity concentrated technologies, available now. And that is part of our project that we are doing in Columbia. There is equipment, its gravity that's brilliant… So, you have a lot of knowledge, as a company and you can actually transfer it to here".—Bruce, vice president of sustainable development at a mine

In collaboration with civil society, mining companies can assist artisanal miners in adopting more sustainable mining techniques. Assistance can be provided by supporting miners with safer mining methods, particularly regarding mercury usage. A mining company, for example, provided training to artisanal miners in Ghana, Burkina Faso, and Indonesia, which included training in mercury-free gold processing techniques (Ribeiro-Duthie & Castilhos, 2016). Such initiatives are also consistent with Action Aid's (2019) recommendation that providing health and safety training and skills development programs is the best way to support the Zama Zamas.

Curbing Exploitation and Extortion

During the research for this book, it was discovered that joint ventures between Zama Zamas and a mining company could solve the exploitation and extortion problem. However, caution is to be practiced when entertaining the domain of cohabitation ideas. While cohabitation can be ideal under certain circumstances, it is important to note that such relationships often have a limited lifespan (Hilson et al., 2020). One conversation took an interesting turn when a mine manager mentioned that the miners might mine outcrops on the mine's concessions. He said this would allow artisanal miners to make a living and reduce the risk of exploitation.

> "For me, this is what I would have done if I was the government. But this, this will only help the people that want to make a living out of this. And not to get rich like the syndicates. But you can, if you can give them some of the sup-outcrops. So a sup-outcrop is the one you can mine from the surface. You can mine it like two or three meters down. Now we do have some of those resources. So if you can give them portions like that and you tell them there's a block of ground, there is some gold here on the surface you can go and mine there".—Will, mine manager

During one of the interviews, a civil society worker voiced the need for a marketplace for the Zama Zamas and said that mining companies ought to convene such a space. Themba said that this would create safer circumstances for the Zama Zamas to trade, provided the miners are paid a fair market price for their gold:

> "…then it allows also the mining house to continue to secure the area and also provide a system how people can, when they come out of underground … can hand over the produce. And then they can be paid safely under those circumstances, provided that they are given a fair market price for their gold".—Themba, civil society

Creating Dialogue for Accurate Information

Another significant finding was the lack of dialogue and accurate information about Zama Zama mining. During the conversations, it was often mentioned that there is little communication between Zama Zamas, mining companies, and state entities, such as the Department of Mineral Resources and Energy (DMRE). Civil society, in particular, expressed an interest in facilitating a dialogue between Zama Zamas and mining companies. A civil society worker described the situation as follows:

> "I think the relations have broken down between the mining houses and the artisanal miners. But certainly, if mining houses were to come to say look we need to be able to discuss with the artisanal miner's certain issues such as how can we better support them, which of the concessions we can give them to mine freely. Then we are in a position to facilitate that kind of discussion".—Themba, civil society

Regarding dialogue, Zama Zamas' inclusion is critical when developing initiatives. During the field research, the Zama Zamas indicated their willingness to meet with mining companies to discuss plans as long as the meetings occur in a safe environment and the miners are not prosecuted. On one occasion, a Zama Zama miner stated:

> "…because there are different people, I can't speak on their behalf, but I am willing to engage".—Sam, Zama Zama

Aside from a lack of dialogue between the two sectors, participants, primarily from civil society, expressed a desire for accurate information to be communicated to the government and the public. Much media coverage portrays Zama Zamas as criminal, illegal immigrant, and violent men (TimesLive, 2018; Fuzile, 2019; Ryan, 2019; Seleka, 2019; Rademeyer, 2022; Mabena, 2023). Representatives from the mining industry paint the same picture. According to the Minerals Council (2018), 70% of all illegal miners arrested are undocumented foreign nationals. Even though no comprehensive statistics on the number and origin of Zama Zamas exist to date, the Minerals Council's document provides concrete figures. During a conversation with a civil society worker, the role advocacy could play in providing accurate information was underlined. Danny commented on the following:

> "Well, I think the first point is advocacy. And advocacy means informing the public, about the real issues and not the emotional issues. Okay. And that means having proper media campaigns around the issues. I think that, that is a form of education for the general public."—Danny, civil society

Conversations between Zama Zama miners, mining companies, and the government can lead to producing accurate information that will balance the narrative around Zama Zama mining.

The Matter of Formalization

During the conversations, some interviewees felt that formalizing Zama Zama mining would solve the risks and challenges within the sector. The following quotations, one from the vice president of sustainable development of a mining company and the other from a civil society worker, reflect this belief:

> "From a company perspective, we have a clear position, all right, where our approach is to look at the formalization of artisanal and Zama Zama mining."—Bruce, vice president of sustainable development at a mine

> "It can all be formalized; it can all be regulated."—Nicky, civil society

Statements such as the above imply that formalization is the silver bullet; however, the interviewees did not specify what this process should entail or what the result would look like. Because so little is known about Zama Zama mining, this fixation on an elusive idea will likely fail. Efforts to formalize (and thus certify) the sector must consider that workers face risks and uncertainties and do not necessarily see regular, salaried employment as a magical solution (Robles et al., 2022). As a result, inclusive formalization efforts should seek to build on what works for all those involved in the sector rather than just a select few.

However, the matter is not so straightforward. During a conversation the complexity of the situation and the mine's role in the matter were outlined.

> "It's a wicked problem and there aren't easy solutions to this stuff. But if you've got all the stakeholders like you say, a multi-stakeholder approach, which where you've got government taking ownership of this thing, you've got civil society providing support, you've got business providing technical support, then that's its possible. To me, it is totally possible. Is there a successful artisanal and small-scale mining formalization process anywhere in the world today to date? My answer to you would be immediately, not that I know of."—Bruce, vice president of sustainable development at a mine

In a discussion with another key informant, the complexity was further unpacked, and the key informant took a somewhat skeptical stance because of the levels of syndicates involved in the matter. She said:

> "We have discussed this a number of times. Formalizing illegal mining would be ideal, but is it possible? Where do you start? Remember, this is a big syndicate that involves communities. It involves industries, companies, it involves officials, police officials, it involves government, you know, so where do you really start? It would be ideal."—Tessa, key informant

In the same vein, another key informant said the government is not really willing to legalize artisanal mining.

> "So really, it's ideal, it's something that would be great. It would develop and increase our economy. It's possible you know it is possible I would say, but you know our officials."—Patricia, key informant

Another important consideration is that formalization will not occur overnight, as seen with South Africa's ASM policy that has been up for implementation since March 2022. Furthermore, even if the policy is formally adopted in the near future, the informal market is unlikely to swiftly change its established practices. This is especially true for artisanal miners, who rely on this livelihood in a country where the unemployment rate is approximately 33.5% as of 2024 (StatsSA, 2024).

It has been suggested (see Chap. 7) that formalization should be viewed as a process that requires ongoing support rather than a onetime event accomplished by simply changing legislation. Scholars have suggested that the best way to achieve formalization is to accompany it with a matching developmental approach (Fourie, 2018). This implies that formalization should support improved livelihoods and more substantial informal enterprises, not as a policy accomplishment.

In line with this suggestion, Webster (2011) proposed that decent work should be an objective to aspire to that can be progressively achieved over several years rather than an immediate goal (Webster, 2011). As mentioned in Chap. 7, the International Labour Organization (ILO) adopted Recommendation 204 in 2015, which concerns the transition from informal to formal employment and includes critical provisions for those informally employed. Recommendation 204 revolves around a call for the government to create a welcoming environment for informal workers and businesses (Chen, 2018). The government is also encouraged to consult with those involved in the informal economy to discuss the transition to the formal economy (Chen, 2018). This demonstrates a much more inclusive approach that encourages the growth of the informal sector and an alternative approach to formalization rather than simply attempting to register those in the informal sector to tax them.

Formalization of artisanal mining has failed on numerous occasions in sub-Saharan Africa, demonstrating that it is not a panacea. This is not to say that some form of regulation is not required; it is, for various reasons, one of which is safety.

However, South Africa is one of the world's most economically unequal countries, so many South Africans have no access to the formal economy or markets. Thus, the formalization process of artisanal mining may result in another formal, organized sector that remains inaccessible to impoverished South Africans.

Formalizing artisanal mining to facilitate control and tax should certainly not be the prime incentive. What is worrying is that South Africa's ASM policy states that entities (meaning the artisanal miners) will be required to pay royalties and taxes (for example, Value Added Tax) to the government, subject to various incentives established by the National Treasury in favor of small businesses.

Additionally, the ASM policy sets strict reporting requirements to monitor the ASM industry's advancement (see Chap. 7). These requirements, as demonstrated by formalization policies in other countries, are likely to perpetuate the illegal and informal status of artisanal miners because of the bureaucratic, technical, and financial challenges they encounter. Therefore, it is crucial that formalization efforts be integrated with developmental initiatives.

The Importance of the Informal Economy

The matter of the informal sector surrounding many Zama Zama sites was raised during conversations with the interviewees. Some mentioned the importance of the informal sector and the need to accept artisanal mining as part of the informal economy rather than trying to formalize it by regulating it. An interviewee indicated that instead of formalizing the artisanal mining sector, it should be embraced for what it is. The account reads as follows:

> "That is my problem. I think if we changed the shift to realize that these are informal, this is an informal economy, and they are here to stay. The one big discourse around the informal economy, and I think that has changed, there was a big hype around we need to formalize, we need to get them to formalize, and we got to a point where we realized, but hang on, but the informal economy is actually here to stay. It's not going anywhere. It employs up to 70 percent of the population in the entire continent".—Jack, key informant.

On another occasion, the following was stated:

> "Okay, so the community gains. They sell food; they sell alcohol. There is, of course, the sex workers, and I don't see a disadvantage. And in fact, if you speak to local people they will say what if we didn't have the gold here we would have people coming, but they would turn to crime."—Carina, key informant

The preceding quotation suggests a larger informal economy centered on artisanal mining. This intertwined web of informal economic activities should be sufficient motivation to begin exploring ways to develop the sector and stop criminalizing it. An entire informal industry will collapse if artisanal mining is constantly restricted and marginalized. Furthermore, scholars have argued that viewing formality and informality as opposites is problematic and misleading (Groenewald et al., 2013).

Because South Africa's formal mining sector is declining, creating a supportive environment for artisanal mining in the informal economy is critical. The lack of support for the artisanal mining sector is a critical oversight regarding socioeconomic development. Artisanal mining can be supported and developed sustainably if decriminalized and allowed to flourish in a nonviolent space in the informal sector.

Current strategies, including criminalizing and policing, are driven by the fact that artisanal miners are operating outside the formal sector and within the informal economy. Such approaches by the formal sector further marginalize the informal sector. Instead of focusing solely on the sector's informality, formal sectors should look for ways to improve the socioeconomic status of the miners.

However, it is never easy for any formal sector to engage with the informal sector. But, as the ILO recognized more than two decades ago, today's flexible labor markets and worldwide production systems make it difficult for corporations to ignore the informal sector (ILO, 2002).

The ILO's recognition echoes the growing acknowledgment amongst mines and the government to follow a more holistic approach to address artisanal mining's challenges and opportunities (SAHRC, 2015). Nevertheless, little has been done to address and develop the sector in the South African context. Corporations face challenges in connecting and dealing with individuals in the informal sector because those workers are mostly not registered and operate illegally or with a dubious legal status (ILO, 2002). Balancing the needs of communities can become tricky. Not only will community members have different needs, but some of their needs might even speak to the issues in the informal sector.

To go a step further, the formal sector faces a challenge because the matters with which the informal sector needs support differ from the needs of large formal enterprises (ILO, 2002). The formal sector might not have the knowledge and skills to deal with informal matters, but ignoring the informal sector is not the best option.

Perhaps one way to bridge the seeming divide between these two sectors is to look at research done by Groenewald et al. (2013). Their research focuses on studies conducted in informal settlements, but it can be applied more broadly and to the context of this research. Groenewald et al. (2013) stated that informal and formal processes are neither separate nor inseparable. Webster (2005) stated that informal and formal economies are inextricably linked. The two sectors have no *"Chinese Wall"* (Webster, 2005). The challenge is that those *"making a living,"* specifically those in the informal economy, are in a lower position within the economy and lack access to economic resources and market data rather than belonging to *"a separate economy"* (Webster, 2005). The challenge for development is to shift away from binary thinking and toward thinking that recognizes the interconnectedness of the informal and formal economies (Groenewald et al., 2013).

Supporting the informal sector becomes urgent in countries with high unemployment rates, such as South Africa. National policies like the National Development Plan focus on the formal sector for economic growth and employment and entirely ignore the informal sector (Fourie, 2018). This is unfortunate since the informal sector holds the potential to offer an income, as can be seen with artisanal mining.

In South Africa, the high levels of unemployment and a declining formal economy have increased the informal economy's growth (Cohen & Moodley, 2012). As mentioned before, South Africa has one of the world's highest unemployment rates, with an official unemployment rate of 33.5% in the second quarter of 2024 (StatsSA, 2024). Additionally, the same quarter of 2024 showed a provincial unemployment rate for Gauteng at 35.1% (StatsSA, 2024).

Therefore, finding alternative ways to make a living is much needed. These alarmingly high numbers reflect South Africa's labor force's dire situation and the fact that job creation initiatives are falling short. Most people enter the informal sector because of limited opportunities in the formal sector (ILO, 2018). Alternative livelihood means are sought within the informal economy because people do not have any other options to make a living otherwise (ILO, 2018).

Recognizing the Sector as an Opportunity for Socioeconomic Development

During the conversations, it was often mentioned that artisanal mining offers a valuable opportunity for socioeconomic development. Many interviewees said that the sector is currently not recognized as a source of socioeconomic development, resulting in a missed opportunity. For example, one of the interviewees had the following to say in terms of economic empowerment:

> "Maybe these artisanal miners, and I know it's not as simple as it sounds, there would need to be proper empowerment and skills development done around this, and proper legislation put in place and, but it's an opportunity for economic empowerment of people who, who are disadvantaged."—Mark, key informant

During a conversation with a civil society member, it was remarked that the high unemployment rate in South Africa and the limited opportunities to create jobs should urge the government to look at the artisanal mining sector as a sector that could create employment. He commented:

> "And, again, when we engage the government, we engage from the fact that [the] unemployment rate in South Africa is almost 30 percent and there is almost no way of creating jobs. So, the government should be looking at these people and say, how can we really make this sector work for the people."—Themba, civil society

The issue of youth unemployment also arose during a discussion with another civil society worker. She said that artisanal mining holds an opportunity to address the country's current youth unemployment rate and a chance to foster growth and redistribute resources.

> "So, we talk about youth unemployment, we talk about all of that, and to not have that middle mechanism, between small-scale sort of company mining and the large-scale industry, leaves a whole lot of gap in between, and I think again, are lost opportunities for employment and growth and … redistribution of resources."—Nicky, civil society

These views advocate for a more inclusive approach to Zama Zama mining. The current approaches, namely criminalization and policing, hinder a socioeconomic opportunity; however, if addressed and supported, the sector offers employment and growth opportunities to those who would not otherwise be able to access it. Furthermore, the sector provides a socioeconomic opportunity for South Africans to access and benefit from the country's resources and wealth. The following quote from a participant captures the way forward and indicates the importance of the sector and proposed ways to address it. A key informant phrased it this way:

> "There needs to be an inclusive discussion held between all parties, which then results in regulatory changes and frameworks that can accommodate such artisanal mining because it is an opportunity for broad-based development within the informal sector. One cannot be narrow-minded in the approaches to economic development opportunities in our nation. So as in other countries in Africa, it provides viable income in impoverished communities."—Mark, key informant

According to research, some of the shafts being worked in by the Zama Zamas are rich enough for small-scale exploitation (Nyoni, 2017). Provided these areas are safe to work in. The Zama Zamas rely on lighter and less expensive equipment and self-employment. These practices could be dormant opportunities to propel local economies (Nyoni, 2017).

Forging a New Path to Empowering Artisanal Miners

Advocating for the sector's socioeconomic development requires a change in perspective. Recognizing the artisanal mining sector's importance and treating it more positively paves the way for decriminalization, offering options beyond criminalization.

This is not to say that it should be an open competition. Regulation is necessary, but it should be implemented in a way that avoids imposing bureaucratic barriers that could restrict access and require resources that artisanal miners lack. According to participants, supporting the sector will contribute to the economic development of other informal sectors linked to the Zama Zama sector. Participants also expressed a desire for dialogue and reliable information; including the Zama Zamas' voices in the development process is critical. This will allow Zama Zamas to develop, add, or reject development programs where civil society plays a significant role. Involving multiple stakeholders, such as civil society and government representatives, in a dialogue will also contribute to accurate and up-to-date information about Zama Zama mining. During discussions with civil society organizations, it became clear that they are eager to facilitate a dialogue between Zama Zamas and mining companies.

A collaboration between mining companies and civil society to launch health and safety campaigns in Zama Zama communities will also improve working conditions in the sector. There is a general need for capacity building among the Zama Zamas. However, training should not be limited to health and safety; general skills development and entrepreneurship training could be significant steps forward (Action Aid, 2019).

The ASM policy that the DMRE published for implementation in March 2022 could be seen, to a degree, as a step in the right direction, even with its pitfalls; however, legislative changes alone will not suffice. Continuous assistance is required during the formalization process, or, as Osumano (2020) suggests, the second step should be to mobilize artisanal miners and assist them in obtaining a mining license. Considering South Africa's ASM policy, exploring and developing meaningful developmental ways to support artisanal miners is vital. Moreover, formalization should focus not on regulation or control but on ways to support informal enterprises (Fourie, 2018). In this sense, formalization should be much more developmentally oriented.

The socioeconomic importance of ASM should be recognized. One way to accomplish this is to integrate ASM activities into a broader livelihood framework by developing supplementary livelihood programs. People should be at the center of socioeconomic development rather than being part of a top-down approach. For the mine to successfully promote sustainable livelihood practices in communities, programs should focus on improving people's well-being. Access to wealth and resources is skewed and divided along racial and class lines in South Africa. As a result, mining companies must consider the needs of surrounding communities and the varying socioeconomic opportunities. This can radically transform local economies, breaking the gold production monopoly held by large-scale operators while empowering communities with local resources (Nyoni, 2017).

It is vital to note the importance of conducting a proper baseline study before any intervention. Interventions are frequently ineffective because they fail to recognize the heterogeneity of artisanal miners (Banchirigah, 2008; Bolay, 2014; Verbrugge & Besmanos, 2016). A comprehensive baseline study of the artisanal mining sector should include all socioeconomic, regulatory, institutional, health, environmental, and technical aspects of all produced minerals (De Haan et al., 2020). A baseline study will also identify women in the sector, their roles, and challenges and allow for developing gender-sensitive initiatives.

Should alternative livelihoods be a consideration to address artisanal mining, it is crucial to consider such solutions within a sustainable livelihood framework. This recognizes that impoverished people may desire more than just money; other tangible or intangible assets may also be required. Creating resilient communities requires a bottom-up approach to development planning that considers local interests, capacities, and strengths. For example, a South African study (Mishi et al., 2020) found that education, internet access, and access to urban areas were all necessary for livelihood diversification (Mishi et al., 2020).

Even so, when it comes to the Zama Zamas, it is crucial to recognize that not all artisanal miners wish to pursue other options. This point was underscored in research by Robles et al. (2022). Their findings confirm that formalization policies must address realities and inequalities in ASM and attempt to account for the preferences of workers, all of whom have different experiences. According to Robles et al. (2022), workers' preferences are not always consistent with mainstream discourses such as the ILO's decent work agenda. It is also important to remember that for some miners, it is their only skill, and the sector's socioeconomic value must not be underestimated.

When it comes to complex issues like artisanal mining, there is no one-size-fits-all solution. Likewise, no macro-level international development policy should be blindly implemented to address Zama Zama mining. Baseline studies and case-by-case research will allow pilot programs to be developed and implemented. Pilot programs and baseline studies should be the foundation of any artisanal mining policy. Such an approach will develop and enforce sensible policies based on the Zama Zamas' and relevant stakeholders' experiences.

A Final Thought

Throughout this book, the voices of artisanal miners and various interviewees have been meticulously presented as direct quotations. These firsthand accounts offer invaluable insights, allowing us to gain a profound understanding of South Africa's artisanal mining sector. Through their everyday experiences, the voices of the Zama Zamas and other interviewees present an opportunity to impact policy and practices supporting South Africa's artisanal miners.

The current state of artisanal mining in South Africa echoes other instances of failed development, but it also offers an opportunity to address injustices truly—both those created prior to 1994 as well as post-1994. What is apparent is that a delicate balance is required when confronting the realities, the drivers, and the potential solutions arising from this matter.

In my work, I frequently encounter mining communities where fundamental human rights are often neglected. In certain South African mining areas, access to essentials like water, sanitation, education, and healthcare remains an unattainable goal for many residents. Thus, addressing these basic needs must come first if we are to truly empower individuals. It is crucial to focus on building personal capacity, as this is the only way to provide impoverished communities with meaningful socioeconomic opportunities.

The importance of poverty alleviation initiatives for historically disadvantaged South Africans cannot be overestimated, and for this reason, there is a great need to strengthen the diversity of livelihoods and alternative economies. This shift in thinking should also incorporate long-term planning rather than the usual once-off arbitrary interventions or programs that mining companies often execute.

South Africa is at a crossroads with the matter of Zama Zama mining. Also, with the ASM policy up for implementation, developmental initiatives are needed more than ever to sustainably support the sector and the miners. Suppose the formalization process is not supported nor coupled with developmental initiatives. In that case, there is a strong likelihood that artisanal miners will be overlooked once more, leaving the sector to operate on the fringes of society and once again face forceful responses from South African authorities.

If artisanal mining is sustainably supported, it could provide, for example, women with a path to full economic empowerment, giving them an alternative to domestic work, as they have explained. It could serve as an economic stepping stone for young people. The possibilities, when the sector is addressed and supported, are endless.

Power has been at the core of South Africa's formal mining industry, often bolstered by unwavering government support. It is long overdue to redefine this power dynamic in mineral access discussions. Empowering marginalized groups, such as artisanal miners, should not mean weakening the power of the government or the formal mining industry. Instead, true empowerment necessitates a fundamental shift toward power sharing. This transformation could create a space where marginalized voices are heard, thereby fostering a more resilient and inclusive social fabric for South Africa.

References

Action Aid. (2019). *Regularising informal artisanal mining in South Africa: An evidence-based report*. Action Aid.

Aubynn, A. (2009). Sustainable solution or a marriage of inconvenience? The coexistence of large-scale mining and artisanal and small-scale mining on the Abosso Goldfields conces-sion in Western Ghana. *Resources Policy, 34*(1), 64–70.

Banchirigah, S. M. (2008). Challenges with eradicating illegal mining in Ghana: A perspective from the grassroots. *Resources Policy, 33*(1), 29–38. https://doi.org/10.1016/j.resourpol.2007.11.001

Bester, V. (2022). A corporate social responsibility conceptual framework to address artisanal gold mining in South Africa. *Resources Policy, 79*, 103030. https://doi.org/10.1016/j.resourpol.2022.103030

Bester, V. (2023). Towards a sustainable artisanal gold mining sector in South Africa: Proposed developmental initiatives. *Journal of Rural Studies, 97*, 375–384. https://doi.org/10.1016/j.jrurstud.2022.12.029

Bolay, M. (2014). When miners become "foreigners": Competing categorizations within gold mining spaces in Guinea. *Resources Policy, 40*, 117–127. https://doi.org/10.1016/j.resourpol.2014.02.001

Calderon, A., Harris, J. D., & Kirsch, P. A. (2016). Health interventions used by major resource companies operating in Colombia. *Resource Policy, 47*, 187–197.

Chen, M. (2018). The South African informal sector in international comparative perspective: Theories, data and policies. In F. Fourie (Ed.), *The South African sector: Creating jobs, informal reducing poverty* (1st ed., pp. 26–47). Human Science Research Council.

Cohen, T., & Moodley, L. (2012). *Achieving decent work in South Africa?* Accessed June 20, 2019, from http://www.scielo.org.za/scielo.php?script=sci_arttext&pid=S1727-37812012000200013

De Haan, J., Dales, K., & McQuilken, J. (2020). *Mapping artisanal and small-scale mining to the sustainable development goals. Minerals, Materials and Society program in partnership with Pact.* Accessed January 20, 2022, from https://www.pactworld.org/library/mapping-artisanal-and-small-scale-mining-sustainable-development-goals

Fourie, F. (2018). Enabling the forgotten sector: Informal sector realities, policy approaches and formalisation in South Africa. In F. Fourie (Ed.), *The South African sector: Creating jobs, informal reducing poverty* (pp. 439–476). Human Sciences Research Council.

Fuzile, B. (2019, March 20). Underworld: Zama-Zamas hold jobseekers hostage. *TimesLive*. Underworld: Zama-zamas hold jobseekers hostage (timeslive.co.za)

Groenewald, L., Huchzermeyer, M., Kornienko, K., Tredoux, M., Rubin, M., & Raposo, I. M. (2013). Breaking down the binary: Meanings of informal settlement in southern African cities. In S. Bekker & L. Fourchard (Eds.), *Governing cities in Africa* (1st ed.). HSRC Press. https://scholar.sun.ac.za/bitstream/10019.1/90428/1/groenewald_breaking_2013.pdf

Hilson, G., Sauerwein, T., & Owen, J. (2020). Large and artisanal scale mine development: The case for autonomous co-existence. *World Development, 130*, 104919. https://doi.org/10.1016/j.worlddev.2020.104919

Intergovernmental Forum (IGF). (2017). *IGF guidance for governments. managing artisanal and small-scale mining*. Accessed February 10, 2017, from http://www.iisd.org/sites/default/files/publications/igf-guidance-for-governments-asm_0.pdf

International Institute for Environment and Development (IIED). (2015). *ASM – LSM – Government relations*. Accessed February 22, 2017, from https://iied.org/sites/default/files/pdfs/migrate/G03938.pdf

International Labour Organisation (ILO). (2002). *Decent work and the informal economy*. Accessed October 3, 2023, from https://www.ilo.org/public/english/standards/relm/ilc/ilc90/pdf/rep-vi.pdf

International Labour Organisation (ILO). (2018). *Women and men in the informal economy: A statistical picture*. Accessed June 28, 2019, from https://www.ilo.org/wcmsp5/groups/public/%2D%2D-dgreports/%2D%2D-dcomm/documents/publication/wcms_626831.pdf

Mabena, S. (2023). Zama zamas spread their roots as illegal mining moves across provinces. *The Citizen*. Accessed October 30, 2024, from https://www.citizen.co.za/news/south-africa/zama-zamas-spread-their-roots-as-illegal-mining-moves-across-provinces/

Martin. (2019). Uncovered: The dark world of the Zama Zamas. *ENACT Africa*. https://enactafrica.org/research/policy-briefs/uncovered-the-dark-world-of-the-zama-zamas

Minerals Council. (2018). *Zama Zama mining fact sheet*. Accessed November 2, 2018, from https://www.mineralscouncil.org.za/industry-news/publications/fact-sheets

Mishi, S., Sikhunyana, Z., Ngonyama, N., & Sibanda, K. (2020). Livelihood strategies and diversification amongst the poor: Evidence from South African household surveys. *The Journal for Transdisciplinary Research in Southern Africa, 16*(1). https://doi.org/10.4102/td.v16i1.726

Nyoni, P. (2017). Unsung heroes?: An anthropological approach into the experiences of 'Zamazamas' in Johannesburg, South Africa. In A. Nhemachena & T. V. Warikandwa (Eds.), *Mining Africa. Law, environment, society and politics in historical and multidisciplinary perspectives* (pp. 133–154). https://doi.org/10.2307/j.ctvh9vz92.7

Osumanu, I. K. (2020). Small-scale mining and livelihood dynamics in north-eastern Ghana: Sustaining rural livelihoods in a changing environment. *Progress in Development Studies, 20*(3), 208–222. https://doi.org/10.1177/1464993420934223

Rademeyer, J. (2022, August 7). Zama zamas just one part of a multibillion-rand organised crime economy that threatens SA. *Daily Maverick*. Accessed October 30, 2024, from https://www.dailymaverick.co.za/article/2022-08-07-zama-zamas-just-one-part-of-a-multibillion-rand-organised-crime-economy-that-threatens-sa/

Ribeiro-Duthie, A. C., & Castilhos, Z. A. (2016). *Brazilian gold mining industry and relations amid ASM, MSM and LSM: Sustainable development initiatives*. Accessed March 1, 2016, from https://www.scribd.com/document/392129562/Brazilian-Gold-Mining-Industry-and-Relations-Amid-ASM-MSM-and-LSM-Sustainable-Development-Initiatives

Robles, M. E., Verbrugge, B., & Geenen, S. (2022). Does formalization make a difference in artisanal and small-scale gold mining (ASGM)? Insights from the Philippines. *The Extractive Industries and Society, 10*, 101078. https://doi.org/10.1016/j.exis.2022.101078

Ryan, C. (2019, March 6). Cable theft and illegal mining threaten Joburg's infrastructure. *Moneyweb.* https://www.moneyweb.co.za/mineweb/cable-theft-and-illegal-mining-threaten-joburgs-infrastructure/.

Seleka, N. (2019, January 9). 6 Zama Zamas arrested during shootout between rival gangs. *News24.* Accessed March 23, 2021, from 6 zama zamas arrested during shootout between rival gangs | News24

South African Human Rights Commission (SAHRC). (2015). *Report of the SAHRC investigative hearing issues and challenges in relation to unregulated artisanal underground and surface mining activities in South Africa.* Accessed November 12, 2015, from https://www.sahrc.org.za/home/21/files/Unregulated%20Artisanal%20Underground%20and%20Surface%20Mining%20Activities%20electronic%20version.pdf

StatsSA. (2024). *Quarterly Labour Force (QLFS) Q2: 2024.* Accessed October 1, 2024, from https://www.statssa.gov.za/publications/P0211/Presentation%20QLFS%20Q2%202024.pdf

Timeslive. (2018, May 9). Zama Zama gold processing hub shut down. *TimesLIVE.* https://www.timeslive.co.za/news/south-africa/2018-05-09-zama-zama-gold-processing-hub-shut-down/

United Nations. (2023). *A call for action for a better world: Nelson Mandela.* Accessed November 15, 2024, from https://southafrica.un.org/sites/default/files/2023-07/NMF%20UNIC%20WEB_1.pdf

Verbrugge, B., & Besmanos, B. (2016). Formalizing artisanal and small-scale mining: Whither the workforce? *Resources Policy, 47*, 134–141.

Webster, E. (2005). Making a living, earning a living: Work and employment in Southern Africa. *International Political Science Review / Revue Internationale de Science Politique, 26*(1), 55–71. http://www.jstor.org/stable/1601650

Webster, E. (2011). The wages are low, but they are better than nothing: The dilemma of decent work and job creation in South Africa. In J. Daniel, P. Naidoo, D. Pillay, & R. Southall (Eds.), *New South African Review 2: New paths, old compromises?* Wits University Press.

Zvarivadza, T. (2018). Artisanal and small-scale mining as a challenge and possible contributor to sustainable development. *Resources Policy, 56*, 49–58. https://doi.org/10.1016/j.resourpol.2018.01.009

Index

V. Bester, *The Untold Story of Zama Zama Miners in South Africa*, Clinical Sociology: Research and Practice, https://doi.org/10.1007/978-3-031-82783-9

R

S

T

U

V

W

Z

The manufacturer's authorised representative in the EU is Springer Nature Customer Service Centre GmbH, Europaplatz 3, 69115 Heidelberg, Germany. If you have any concerns regarding our products, please contact ProductSafety@springernature.com

Printed and bound by CPI Group (UK) Ltd, Croydon, CR0 4YY
01/10/2026
02225977-0017